DISCARD

THE POWER OF THE GENE

THE ORIGIN AND IMPACT OF GENETIC DISORDERS

GENETICS – RESEARCH AND ISSUES

Additional books in this series can be found on Nova's website
under the Series tabs.

Additional E-books in this series can be found on Nova's website
under the E-book tabs.

GENETICS – RESEARCH AND ISSUES

THE POWER OF THE GENE

THE ORIGIN AND IMPACT OF GENETIC DISORDERS

CHRIS MURGATROYD

Nova
Nova Science Publishers, Inc.
New York

LIBRARY OF CONGRESS CATALOGING-IN-PUBLICATION DATA
Murgatroyd, Chris.
  The power of the gene : the origin and impact of genetic disorders / Chris Murgatroyd.
    p. ; cm.
  Includes index.
  ISBN 978-1-60876-949-0 (hardcover)
  1. Genetic disorders. I. Title.
  [DNLM: 1. Genetic Diseases, Inborn--physiopathology. 2. Cell Physiological Phenomena. 3. DNA Damage. 4. Genetic Diseases, Inborn--genetics. 5. Genetic Predisposition to Disease. QZ 50 M976p 2009] RB155.5.M87 2009
  616'.042--dc22
                            2009052746

Published by Nova Science Publishers, Inc. + New York

CONTENTS

Preface   vii

Introduction   1

Chapter 1   Cell Division and Chromosome Defects   9

Chapter 2   Patterns of Inheritance   21

Chapter 3   Skeletal Disorders   39

Chapter 4   Connective Tissue Disorders   55

Chapter 5   Muscular Disorders   59

Chapter 6   Dermatological Disorders   67

Chapter 7   Hair Disorders   83

Chapter 8   Respiratory and Heart Disorders   91

Chapter 9   Blood Disorders   99

Chapter 10   Immune Disorders   115

Chapter 11   Endocrine Disorders   123

Chapter 12   Metabolic Disorders   137

Chapter 13   Renal, Liver and Digestive Disorders   145

Chapter 14   Visual Disorders   153

Chapter 15   Hearing Disorders   165

Chapter 17   Neurological Disorders   171

Chapter 16   Cancer and Disorders of DNA Repair   187

Glossary   195

Index   199

PREFACE

Our genes are the hand we are dealt at birth. Unlike poker we do not have opportunity to swap unfavourable genes during the game of life, though research and advances in biotechnology are just beginning to change the rules in this regard. This book explores how genes, and in particularly genetic disorders, have impacted on peoples' lives. Some people are dealt marked cards and when this happens to influential people the impact of the disorders goes far beyond individual lives. Through their impact on kings and queens, and on political leaders, genetic disorders have altered the course of World history. Historical events that have been influenced directly by genetic disorders include, for example, the Russian Revolution and the American War of Independence. Through their impact on great painters such as Van Gogh and great composers such as Beethoven, genetic disorders have guided the evolution of art and music. Even much of the myth and folklore that has come down to us through the ages has been fashioned by genetic disorders. Examples include the tales of elves, vampires and werewolves. The many stories in the book are used to draw the reader through the principles of genetics towards an understanding of the origin and inheritance of genetic disorders and an appreciation of the power they wield. Many of the examples used relate to historical figures and celebrities but the vast majority of the lives devastated by the random cruelty of genetic mutations are otherwise ordinary ones. But genetic disorders can turn ordinary lives into extraordinary ones. So often in these cases the power of the human spirit soars above adversity and many of those so cruelly affected have fought back in their efforts to expand awareness of genetic disorders and thereby raise funds for research. This book is respectfully dedicated to those people, and half of all the royalties accrued by this book will go towards this research - research that is now just starting to rein back the power of the gene.

INTRODUCTION

Although there is no universal agreement as to a definition of life, its biological manifestations are generally considered to be organization, metabolism, growth, irritability, adaptation, and reproduction.

The Columbia Encyclopaedia, First Edition, first sentence of the article on "life", 1935

When in 1665 Robert Hooke examined a thin section of cork through a crude homemade microscope our understanding of life science changed forever. The rows of tiny boxes which made up the dead wood's tissue reminded him of the rows of cells occupied by monks in a monastery; and so the word cell passed into scientific terminology. Around the same time the Dutchman, Antoni van Leeuwenhoek, became the first to describe living cells including bacteria and sperm. Two centuries were to pass, however, before scientists fully grasped the true importance of cells.

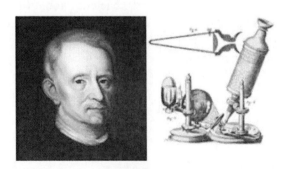

Robert Hooke.

We now know that a cell is the smallest structure capable of basic life processes, such as absorbing nutrients, expelling waste, and reproducing. Some microscopic organisms such as bacteria and protozoa consist of a single cell. Multicellular organisms, however, generally consist of several different varieties of cell each designed to perform different tasks within the organism. These different kinds of cell working in concert allow the organism to function as a single unit. The human body is made up of around 20-30 trillion cells; if these were placed end to end they would stretch around the earth 47 times! There are over 200 different varieties varying in size from the female gamete (also known as the egg, oocyte or ovum) which is about one millimetre (1,000 μm) in diameter - just visible to the naked eye - to the smallest human cell, the male sperm, that is just 60 μm diameter. Not all cells are compact and a great variation in shape exists with the longest being nerve cells (also known as neurons) connecting the end of the toes to the spinal cord and so spanning half the human body. A much quoted contender for the longest cell in the animal kingdom are the nerve cells running down a giraffe's neck which can be as long as 3 meters!

All cells that make up multicellular organisms (called eukaryotic) have a nucleus containing DNA (deoxyribonucleic acid). This is in contrast to the simpler prokaryotic cells, found only in bacteria, where all the components, including the DNA, mingle freely inside the cell's interior. DNA is the hereditary material in all living organisms (except for some viruses that use RNA - ribonucleic acid), providing the coded instructions for the production of proteins which in turn regulate the operation of the organism as a whole.

To gain an understanding of how DNA, proteins and cells fit together one needs to look back through the evolutionary story of life on Earth to its very beginning. Our planet formed around 4.6 billion years ago, and for millions of years, volcanic eruptions pumped gases such as carbon dioxide, nitrogen and water vapour into the atmosphere. It is likely that the combined effects of ultraviolet radiation, lightning from intense storms and great variables in temperature on our early atmosphere led to the synthesis of much larger molecules such as amino acids and nucleotides—the respective building blocks of proteins and nucleic acids (DNA and RNA). Some laboratory experiments involving simulating conditions similar to the early environment of Earth have provided evidence for this idea, although as yet, it is still only a hypothesis. It has also been demonstrated in laboratory experiments that lipid (fat) molecules can join to form spheres resembling a cell's membrane and so it seems reasonable to speculate that millions of years of molecular collisions resulted in lipid spheres enclosing RNA, the simplest molecule capable of self-replication. These primitive aggregations would have been the ancestors of the first prokaryotic cells, evolving around 3.5 billion years ago. As there was no oxygen at this time these early bacteria would have used fermentation to produce energy. These then evolved to perform photosynthesis, producing oxygen as a by-product, with the result that there was a gradual accumulation of oxygen in the atmosphere. This then set the stage for the evolution of bacteria able to use oxygen in aerobic respiration, a more energy-producing process than fermentation.

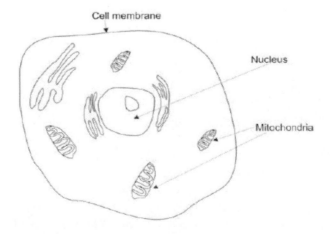

A simplified diagram of an animal cell.

Eukaryotic cells, in addition to containing a nucleus, also contain other organelles (a specialized subunit in a cell like an organ in a body) such as mitochondria (in animal cells) and chloroplasts (in plant cells). These eukaryotic cells evolved from the more primitive prokaryotic cells around 2 billion years ago, with one widely accepted hypothesis being that the mitochondria and plant chloroplasts were once free-living bacteria that were engulfed and maintained inside other cells for their ability to produce energy. This allowed the host cells, which previously relied on fermentation, the ability to conduct cellular respiration. This greatly expanded the number of environments in which the cells could survive and so played a critical part in the development of eukaryotic cells and the consequent evolution of multicellular organisms.

The so-called blueprint of life, DNA, is the hereditary material in all living organisms and is found in all living cells, apart from some viruses that contain the related RNA. This DNA is usually packaged into structures called chromosomes, which were first observed in plant cells in 1842 (Karl Wilhelm von Nägeli) using a special dye; hence the name chromosome derives from Greek words *chroma* (colour) and *soma* (body). However, it was not until 1910 (Thomas Hunt Morgan) that it was discovered that the chromosomes that actually carried the hereditary material, and it was not until 1944 that this hereditary material was found to be DNA (Avery, MacLeod and McCarty) through a series of experiments revealing that only DNA could enable organisms to produce proteins. It was then 6 years later, in 1950, that James Watson and Francis Crick famously published their model for the molecular structure of DNA, as a double-stranded helix. Their ideas were supported, and possibly even prompted, by X-ray photographs of DNA carried out by a scientist called Rosalind Franklin, who worked at Kings College London. Unbeknown to her, another researcher at the university, by the name of Maurice Wilkins, had secretly showed Watson and Crick the photographs which seemed to suggest the helical structure. Soon afterwards they published their hypothesis in the journal Nature, and in 1962 Watson, Crick and Wilkins collected the Nobel Prize. By this time Rosalind had died, in relative obscurity, of ovarian cancer.

Rosalind Franklin.

Chemically speaking, DNA is a long molecule made up of a sequence of four chemical bases: adenine (A), guanine (G), cytosine (C), and thymine (T), which pair up with each other (A with T and C with G) to form units called base pairs. Each base is also attached to a sugar and a phosphate molecule, which together are arranged in two long strands forming a spiral called a double helix. This structure can be thought of as a ladder, with the base pairs forming the ladder's rungs and the sugar and phosphate molecules forming the vertical sidepieces. An important property of DNA is that it can replicate, or make copies of itself - each strand of DNA in the double helix can serve as a pattern for duplicating the sequence of bases. This is critical when cells divide because each new cell needs to have an exact copy of the DNA present in the old cell.

The entire DNA present in the nucleus of a cell, contained in all the chromosomes, is referred to as the genome. The human genome consists of about 3 billion of the four different bases forming the DNA, arranged in a specific, yet unique, order to each of us. If we take the ½ million letters in this book, imagining them as individual G, A, T or C bases, then the genome of the Phi-X-174 virus, which infects *E. coli*, would be about twice the size of this chapter (5,386 base pairs). The *E. coli* genome would be a little longer than the size of this book (4,639,221 base pairs), while the fruit fly genome would come to around 35 books (122,653,977 base pairs). The genome of the dog (2.4 x 10^9 base pairs) would take almost 7,000 copies of this book and the human (3.3 x 10^9 base pairs) would take 9,500 books. Interestingly though, the size of the genome often bears little relation to the complexity of the organism - cells of the onion and lily contain respectively five and 30 times as much DNA as a human cell!

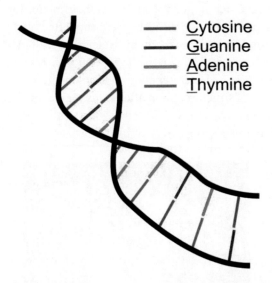

The structure of DNA.

It is amongst these 3 billion DNA base pairs of the human genome that an estimated 20,000 to 25,000 genes are found. Deriving from a term Darwin coined to describe a unit of heredity (Gr. *genos*; origin), the basic - but not perfect - definition of a gene is the region of DNA that can code for a protein. The journey from gene to protein involves the two steps of transcription and translation. During the process of transcription, the information stored in a gene's DNA sequence, is transferred to a similar molecule called RNA (ribonucleic acid). This RNA carries the information out of the nucleus and into the cell for translation where the RNA message is used to assemble amino acids to form a protein. Three bases code for one particular amino acid and there are 20 amino acids which, when arranged in different orders, are the building blocks of all proteins. For example, the protein vasopressin is composed of 9 amino acids in the sequence of Cysteine-Tyrosine-Phenylalanina-Glycine-Asparagine-Cyteines-Proline-Arginine-Glycine. All other proteins are composed in the same way; the largest to date is the aptly named titan, a protein found in muscle, which is composed of 33,000 amino acids!

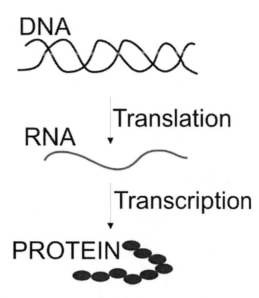

DNA sequence is "transcribed" into RNA which is then "translated" into an amino acid sequence making up a protein.

As an example of a genome, and to give a sense of how genes are organised, the entire sequence of the Zaire Ebola virus is given (note that this retrovirus actually has an RNA genome). First of all it can be seen that some viral genomes are so small as to fit on one page, and secondly the genome contains only 7 genes (shown in the different colours) encoding 7 different proteins. It is incredible to think that something so small can be so deadly, causing the recurrent outbreaks of deadly hemorrhagic fevers, in central Africa, with mortality rates as high as 90%. One of these proteins, for example VP40 (shown in pink), functions to allow newly replicated Ebola viruses to exit cells, often killing those it has invaded, and infecting other host cells until the cumulative effect of this explosive viral proliferation overwhelms the body.

The sequence of the series of amino acids making up this VP40 protein, given below, is coded for by the gene DNA sequence (in pink) which can be read like a simple code book. For example, if we start at the beginning of the gene we have: atg agg cgg gtt…. The first three 'letters' atg always codes for the amino acid methionine (M), followed by 2 arginines (R) that are coded by agg or cgg, then a valine from gtt, and so on.

One particular region highlighted in the protein (PPEY), consisting of two prolines (P), glutamic acid (E) and tyrosine (Y), is especially important for the protein to function properly. Therefore, some groups are looking for cures for Ebola by inactivating this region. As no approved vaccine or treatment is yet available, Ebola is classified at the highest biosafety level possible and incredibly, has been considered for use in biological warfare by both the USA and the former Soviet Union. While the genome of the primitive Ebola virus contains only 7 genes, we humans have considerably more. It is therefore certainly natural to assume that the number of genes correlate with the complexity of an organism. If this were true we would have far more genes than any other creature on the planet. However, this is not the case and recent sequencing has shown that we have considerably less genes than first imagined. The estimate at present gives us around 20,000-25,000. However, this does not solve the puzzle as to why many organisms, seemingly less "advanced" than ourselves, can have more genes than we do. While sharing over 99.9% identical gene sequences among ourselves as a species, regardless of race, we also share about 96% of our genes with chimpanzees. Furthermore we share 80% of our genes sequences with mice, 75% with dogs, 50% with the fruit fly (*Drosophila*) and 30% with yeast; we even have some genes in common with bacteria. One explanation for how this might be is that many genes are capable of making more than one protein, allowing human cells to make perhaps 80,000-100,000 proteins from our 20,000-25,000 genes.

Amino acid sequence of the Ebola VP40 protein

MRRVILPTAPPEYMEAIYPVRSNSTIARGGNSNTGFLTPESVNGDTPSNPLRPIADDTIDHASHTPGSVSSAFILEAMVN
VISGPKVLMKQIPIWLPLGVADQKTYSFDSTTAAIMLASYTITHFGKATNPLVRVNRLGPGIPDHPLRLLRIGNQAFLQEF
VLPPVQLPQYFTFDLTALKLITQPLPAATWTDDTPTGSNGALRPGISFHPKLRPILLPNKSGKKGNSADLTSPEKIQAIMT
SLQDFKIVPIDPTKNIMGIEVPETLVHKLTGKKVTSKNGQPIIPVLLPKYIGLDPVAPGDLTMVITQDCDTCHSPASLPAVI
EK

Chimpanzees share 96% of our DNA.

Taken together, what actually defines us genetically is not e how many genes we have; it is the complexity of how our genes are used, and how these genes are controlled to interact with one another to carry out the various functions, that give us our unique characteristics as humans.

Chapter 1

CELL DIVISION AND CHROMOSOME DEFECTS

For each chromosome contributed by the sperm there is a corresponding chromosome contributed by the egg; there are two of each kind, which together constitutes a pair.

Thomas Hunt Morgan

If the 3 billion base pairs of DNA in a single human cell were stretched end-to-end, it would extend over 2 meters! This is of course many thousands of times longer than the cell in which it is found. To allow this entire DNA to fit into the nucleus of a cell, it is packaged - by twisting, folding and wrapping around proteins - into the much more compact structure of chromosomes. These are around 5 μm long and so represent a 400,000-fold reduction in length!

All animals, in the nucleus of their cells, have a characteristic number of chromosome pairs known as a karyotype (Gr. *Karyon*; nut, used to donate the nucleus). The number of chromosomes in an organism bares little relation to the complexity of the organism: mice have 40 chromosomes, the dog and chicken both have 78, the cells of fruit flies contain 8, some ant species contain only 2 and some crayfish species 200 chromosomes!

It took until 1956, over a century after chromosomes were first observed, before the correct number of chromosomes in humans – 46 - was established. This was actually aided by a laboratory accident with hypotonic saline that swelled the cells allowing the chromosomes to separate sufficiently to get an accurate count. These 46 chromosomes occur as 23 pairs found in each cell of our body. The only exceptions to this rule are the gametes (i.e. eggs and sperm) which carry only one copy of each of the 23 chromosomes - these single copies combine at fertilisation to give a total of 23 pairs again. In 1971, at a conference in Paris, these chromosomes were numerically named from 1 to 22 (the 23rd pair were called 'X' and 'Y') according to their size. This scheme of coding chromosomes is still in use today, though in making the original assessments of size a mistake was made - chromosome 22 is actually longer than chromosome 21 and so the numbers should have been reversed. The first 22 chromosome pairs are referred to as autosomes whilst the other pair, known as the sex chromosomes, determines the sex of the individual; males have an X and a Y chromosome, while females have two X chromosomes.

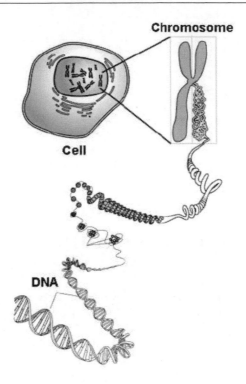

Each cell of an organism contains DNA which is packaged as a chromosome.

Cells multiply through a process of cell division. Two types of cell division occur in eukaryotes: mitosis, and meiosis. Mitosis produces two genetically identical cells from a single cell and is used by multicellular organisms for growth, cell repair, and cell replacement. In the human body, for example, an estimated 25 million cell divisions occur by mitosis every second in order to replace cells. Different cell types divide at different rates and while cells of the liver, intestine, and skin may be replaced every few days, neuronal cells in the brain never divide again after they differentiate - your present neurons are exactly the same age as you!

The mechanism of meiosis, needed for sexual reproduction, is slightly different – this needs to be the case so that we do not end up producing a genetically identical version of ourselves. The way in which this happens is that special oocyte and sperm cells are produced carrying only one copy of each of the 23 chromosomes. When they unite during fertilization, the cell resulting from the union will contain the full 23 pairs of chromosomes again. Consequently, our offspring will contain exactly half our genetic material; only one copy from each of our chromosome pairs, and so only one copy of each of our gene pairs. The resulting offspring will therefore gain 23 pairs of chromosomes with one copy of each gene coming from each parent.

As a cell gets ready to divide, chromosomes are separated into each of the dividing cells by a complex of proteins which bind and separate the chromosomes by physically moving them. A failure in this chromosome distribution during the process of cell division, in the making of gametes, results in sperm or oocytes lacking, or gaining, a chromosome. At fertilization these will produce cells, and subsequent embryos, containing only a single chromosome or three chromosomes (trisomy) instead of the usual pair. These numerical chromosomal abnormalities tend to occur more in females during oogenesis (the production of oocytes), and the chance of this occurrence increases with maternal age. The reason this happens more during oogenesis is due to the different way that eggs and sperm are produced. All of the eggs that are ever going to develop in the female are present at birth and can be more than 40 years old. Therefore, it is suspected that the aged molecular apparatus in these cells, involved in the cell division, leads to mistakes in chromosome separation as a cell divides.

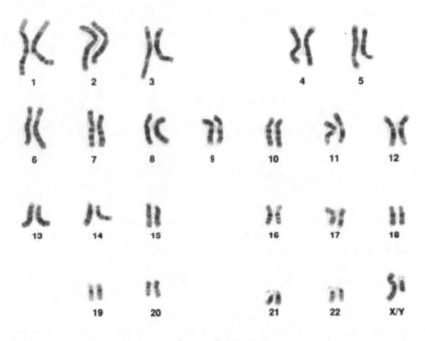

A human karyotype.

Men, on the other hand, produce new sperm continually from puberty throughout lifetime. Though this process generally decreases during old age, males in their 80s have been known to father children. The oldest recorded was Les Colley, a miner from a town in Victoria, Australia, who was nearly 94 when his son Oswald was born in 1992. "*I never thought she would get pregnant so easy, but she bloody well did,*" he told the papers. While there is no increased risk for chromosome abnormalities to occur based on the age of the father, sperm production is more prone to gene mutations, especially with age. Some genetic diseases such as achondroplasia, mentioned in the next chapter, occur more frequently with older fathers.

Of all pregnancies around 15-20 percent end in miscarriage and about half of these have a chromosome abnormality. These either involve foetuses inheriting abnormal numbers of chromosomes or structural defects in the chromosomes. Chromosomal abnormalities can occur in around 1 in 200 live births and account for at least half of all miscarriages that occur during the 1st trimester. Although extra numbers of any chromosome can occur, only extra copies of the X and Y chromosomes, the 21st and, to a smaller extent, the 18th and 13th chromosomes, have any compatibility with life. A foetus lacking any chromosome of a pair, apart from the X chromosome, is unable to survive.

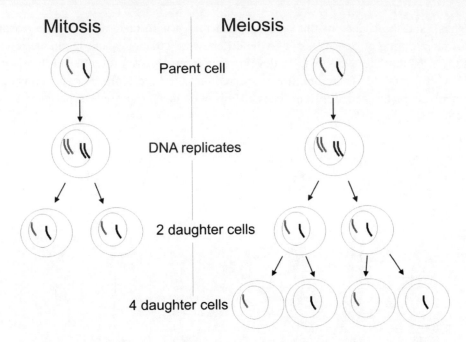

A comparison of mitosis and meiosis. The one orange and one black chromosome constitute a chromosome pair. Mitosis produces two daughter cells that are identical to the parent cell. Meiosis produces daughter cells that have one half of a chromosome pair.

An individual inheriting an extra chromosome 21, thus having three copies of chromosome 21 (known as trisomy 21), has Down syndrome. This extra set of genes on chromosome 21 produce increased levels of their respective protein products, resulting in a number of characteristics. Occurring in around 1 in 700-900 births, this is associated with a number of medical problems including increased risks of hearing and vision defects, heart abnormalities and also a range of developmental difficulties involving delayed coordination skills and mental abilities such as speech and short-term memory.

Chris Burke.

Importantly, the range of abilities and disabilities among people with Down syndrome varies widely, as in the general population, and the majority can lead independent lives. Chris Burke, best known for his role as *Corky Thatcher* in the TV series *"Life Goes On"*, became the first person with Down syndrome to star in a weekly television series. He is now the goodwill ambassador for the National Down Syndrome Society and serves as the editor-in-chief of its magazine, *Straight Talk*, as well as performing and touring with a music group.

Edward syndrome involves trisomy 18 and, like Down syndrome, affects all systems of the body. It is estimated to occur in 1 in 6,000-8,000 births, though around 95% of foetuses die before birth, so the actual incidence of the disorder may be higher. Of those born, the majority are females and die within the first year with symptoms including mental retardation and delayed development with malformations such as brain defects and microcephaly (small head). The presence of an extra chromosome 13 results in the even more severe Patau's syndrome with death usually occurring by 1 year old.

Normal males inherit an X and a Y chromosome while females have two X chromosomes. Because females have two X chromosomes, they must inherit two copies of every gene on the X chromosome. In contrast males only inherit one X chromosome and one Y chromosome. This might mean that females could produce twice the amount of protein from these genes, on the X chromosome, than males. Therefore, to overcome this, females maintain only a single active X chromosome in each cell, while the second X chromosome becomes inactivated. This inactivated X chromosome is referred to as a Barr body and different cells randomly inactivate either of the two X chromosomes, so it is not the same X chromosome activated in every cell. This is in contrast to marsupials, like the kangaroo, where it is always the father's X chromosome that is inactivated in the female. However, there arc around 18 genes on the X chromosome that are also found on the, very gene-sparse, Y chromosome. Therefore, as there should be no gene imbalance between males and females for these genes, they escape inactivation on the otherwise inactivated X chromosome in females.

The role of the Y chromosome in mammalian male sex determination was first confirmed when a male calico cat was found. Ordinarily, calico and tortoise-shell cats are females since the gene for coat colour is found only on the X chromosome. Two variants (known as alleles, which are described later) of this gene exist, one producing black fur and another giving ginger. Therefore, only a female with two X chromosomes and the two different alleles for the black and ginger colours can produce the calico colour patterning. The pattern of a calico cat is a perfect example of X inactivation pattern; the patches of ginger and black depend on which X chromosome is active i.e. if the X chromosome carrying the black colour gene is inactivated, then the skin here will produce the ginger colour from the active X chromosome. Therefore, when a male calico cat was found (who had an XXY sex chromosome karyotype) it showed that the presence of two X chromosomes did not lead to femaleness, and instead confirmed that it is a copy of a Y chromosome that is needed to determine maleness.

The XXY anomaly is also found in humans - known as Klinefelter syndrome where individuals who inherit more than one X chromosome but still possess a Y chromosome are male. Occurring in around 1 or 2 in 1,000 births, making it one of the most common chromosomal abnormalities, Klinefelter syndrome individuals are characteristically slightly taller than average, with normal mental ability, though sometimes showing feminine body contours (which can be treated with testosterone administration during puberty) and little facial and body hair. They generally also have small testes and prostrate glands. While most individuals with the condition are usually capable of normal sexual function, many are unable to produce sufficient amounts of sperm for conception. However, it should also be noted that people with Klinefelter syndrome are not more inclined to be homosexual, although a small number may be transgendered. The artist Lili Elbe was thought to have had this disorder. Born Einar Wegener in 1886, she was raised as a male and became an accomplished artist, though was often assumed to be a young woman in man's clothes. He married Gerda Wegener, one of the most influential Art Deco artists of the early twentieth century, and cross-dressed for her when she needed a female model. He actually became Gerda's favourite model and the 1920's small breasted feminine ideal may have been influenced by Lili's figure. Lili then underwent sex reassignment surgery and it was shortly after one of these surgical procedures that she died in 1931. The story of her life is dramatised in the novel *The Danish Girl* by David Ebershoff and is set to be made into a maovie.

A calico cat.

Some people have inherited more than two X chromosomes along with the Y chromosome resulting in karyotypes such as XXXY, which are still considered forms of Klinefelter syndrome. This is the chromosomal arrangement of Caroline Cossey, also known as Tula, who appeared in the *James Bond* movie "*The Spy who Loved Me*". Though raised as a boy, she opted to live as a girl at a young age, undergoing sex-assignment surgery and subsequently becoming a well known model and actress.

The inheritance of more than two X chromosomes in the absence of a Y chromosome results in a so called metafemale, or triple-X female. These women can possess three or more X chromosomes with a genotype of XXX or more rarely XXXX or XXXXX. Remember that in females with two X chromosomes, one of the X chromosomes is inactivated. However, also recall from the previous page that there are a small number of genes that still remain active and so, while normal females (i.e. XX) still have two functional copies of some genes on the X chromosome, individuals with multiple X chromosomes consequently have more functional copies of these few genes. These genes code for proteins important in bone growth, in addition to some other systems, explaining why individuals with this disorder are consequently taller than average with unusually long legs and slender torsos, but otherwise appear normal and are fertile. This type of chromosomal abnormality occurs in around 1 in 1,000 females and, in common with all numerical chromosome abnormalities, occurs more often when the mother is older. One lady with this anomaly was Ewa Klobukowska, a 1964 Olympic sprint bronze medallist who had the dubious honour of being the first woman to fail the sex chromosome test during the 1967 European Cup. *"One chromosome too many to be declared a woman for the purposes of athletic competition"* was the statement in response to finding her to be XXX, and she was obliged to return all her medals. *"I know what I am and I know how I feel"* she said at the time, and made a mockery out of the whole investigation by giving birth to a baby the following year.

Les deux amies (possibly Gerda and Lili). Gerda Gottlieb Wegener, 1921.

Ewa Klobukowska, 1964.

Originally, the sex test introduced by the Olympics Committee required women to disrobe for medical staff leading to early retirements of some athletes such as the famously masculine Russian shot putters the Press sisters. However, developments in reconstructive surgery to remove external male-like reproductive structures lead to the introduction of the sex chromosome test - any female not having exactly two X chromosomes was disqualified.

Between 1972 and 1990, one in 504 female athletes was found to be ineligible for female events at the Olympics as a result of this chromosome test. However, none was found to be a physically normal male, with the majority found to suffer from androgen insensitivity syndrome (described in the chapter on *Endocrine Disorders*) which, though characterised by a male XY genotype, actually results in a completely female physique. Erika Schinegger, who won the Olympic gold medal for Austria for women's downhill skiing in 1966, failed the sex chromosome test. After being diagnosed with this condition, she went on to father a child. The sex chromosome test was finally abandoned in 1990 in favour of physical examinations and, although in the Atlanta 1996 Olympic Games around 1 in 400 of the female competitors were found to have chromosomal abnormalities, all of them subsequently passed physical examinations.

In contrast to the inheritance of multiple X chromosomes, extra Y chromosomes can occur in an individual leading to what is known as XYY syndrome. As adults, these men are usually taller than average and generally appear normal. However, they produce high levels of testosterone due to the dosage effect of the extra Y chromosome which, during adolescence, can lead to severe facial acne. Occurring in around 1 in 1,000 males, the majority lead normal lives unaware of their chromosomal abnormality. Though these individuals are usually fertile, a percentage of people with XYY syndrome are infertile, due to the production of increased amounts of certain hormones leading to inadequate sperm production. Stefan Kiszko, who was jailed for the murder of 11-year-old Lesley Molseed in Manchester in 1975, had XYY syndrome. He spent 16 years in prison before he was released, after evidence showed that his semen samples contained no sperm, in contrast to the semen found on Lesley Molseed's clothes, which did contain sperm. The semen sample from the crime scene was subsequently used to provide the DNA evidence which helped to convict the real murderer who is currently serving a life sentence for the crime. However, it is still often suggested that the high testosterone levels of XYY men can make them more prone to violence and criminal activity. Much of this stems from some studies between 1965 and 1968 although many later studies cast serious doubt on any direct and simple linkage. Around this time was the murder trial for Richard Speck who raped and brutally murdered 8 nurses in a Chicago dormitory. His attorney argued that his acne and aggressive behaviour was proof of XYY and therefore he could not be considered legally responsible for his uncontrollable urges. Two separate chromosome tests that showed him to be normal XY were overlooked in the confusion of courtroom drama, and decades later many books still associate Speck with the XYY condition. Nevertheless he was sentenced to life in prison where he eventually died. The idea of a link between XYY and criminality was used as the basis of the television fiction series "*The XYY man*" about an individual with XYY, and thereby a natural criminal. The "*Alien 3*" movie also used this theme, being set in an off world penal colony for XYYs.

While a small percentage of individuals inherit more than the normal number of two chromosomes, the inheritance of just a single chromosome of a pair can only occur with the X chromosome. This is known as Turner syndrome and affects around 1 in 2,500 females. In contrast to metafemales these individuals lack the extra functional copies of those genes on the additional X chromosome and are therefore usually short in stature, averaging 140 cm. While most women with Turner syndrome have underdeveloped ovaries and do not ovulate, regular growth hormones and oestrogen replacement therapy can increase height and allow menstruation to occur.

While mistakes can happen during meiosis in the production of sperm and oocytes, that can lead to a foetus developing with different numbers of chromosomes, this can also occur during mitosis in cell division in other cells of the body. **Some diseases, such as Robert's** syndrome, present with a high occurrence of this type of defect due to mutations in a gene producing a protein important in moving chromosomes into each of the two dividing cells. **As a result, an individual's cells do not divide properly or stop growing** leaving such sufferers with shortened limbs that resemble those of babies whose mothers took thalidomide during pregnancy. For this reason this syndrome is also referred to as pseudothalidomide. It should be noted though that thalidomide causes birth defects through completely different mechanisms, possibly mainly involving disruption to the growth of blood vessels.

It has been suggested that Roberts syndrome might have been behind the legend of the *Monster of Ravenna*. This was based on the birth of a severely deformed child in 1512 in the Italian city of Ravenna. Shortly afterwards Italian forces were defeated in the Battle of Ravenna. Though the infant only survived a few weeks, the birth was taken as an evil omen and word of mouth progressively exaggerated the deformities – the rumors leaving Florence were of a the child with two serpentine legs and by Paris it had a single claw.

In contrast to abnormal numbers of chromosomes, structural defects of chromosomes can also occur. This can involve a loss of a piece of chromosome (known as a deletion), or else a duplication or a rearrangement in the location (i.e. translocation) of part of a chromosome where, for instance a piece of chromosome reattaches itself in a different part of another chromosome. Depending on which chromosome region is lost or altered, and which genes are subsequently affected, specific disorders are seen.

One example of a deletion is Williams syndrome, occurring in around 1 in 20,000 people, where a small part of chromosome 7 is missing. This region contains a number of genes which, when lost, lead to cardiovascular problems and a generally low IQ. However, these individuals tend to show competence in areas such as language and music, with many showing a near perfect music pitch and an uncanny sense of rhythm. People with Williams **syndrome also have characteristic 'elfin-like' facial features with a small upturned nose,** depressed nasal bridge and a broad mouth with full lips. This, together with their often remarkable musical and verbal abilities, and highly sociable depositions, has lead to the suggestion that affected children were the inspiration for folktales and legends such as pixies, elves and fairies, who were often musicians and storytellers.

Unrelated children with Williams syndrome. An illustration of an elf by the artist Richard Doyle.

FLUORESCENT IN SITU HYBRIDIZATION (FISH)

FISH is a method to allow for the detection of small chromosome aberrations. The chromosomal deletion causing Williams Syndrome is so small that it cannot be seen with a microscope. However, the deletion can be observed using a special technique called fluorescent in situ hybridization, or FISH. In this technique, a fluorescent chemical is attached to a synthetic piece of DNA, called a probe, which can recognize specific DNA sequences.

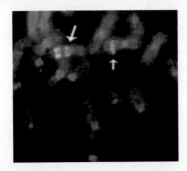

The normal chromosome (left) shows a green signal (control for chromosome 7) and a pink signal (Williams syndrome region) while an abnormal chromosome (right) contains only a green signal, i.e. the Williams syndrome region is missing.

The probe then binds to the matching DNA sequence on the chromosomes and fluoresces highlighting the location of the probe-bound DNA sequence. If fluorescence from the probe occurs, the gene is present. - if not, the gene has been deleted. The Williams Syndrome deletion can be detected by labelling a piece of chromosome 7, normally absent in Williams syndrome, with a fluorescent probe. This can be combined with a probe labelling another area of chromosome 7. Therefore a signal for chromosome 7 in the absence of a signal for the deleted region will signify Williams syndrme.

In contrast to chromosome deletions, it is possible for a part of a chromosome to become duplicated. An example of this occurs in the rare Cat-eye syndrome in which there is a duplication of part of chromosome 22. Although a highly variable condition some sufferers are missing tissue from the eyes called, a defect known as a coloboma, which can give the pupils a catlike appearance. Occurring in around 1 in 10,000 in the population, this defect is a distinguishing feature in the right eye of Madeline McCann, the little British girl abducted in Portugal in May 2007. Due to the relative rarity of this anomaly, the word "*LOOK*" with the first "*O*" in the word being drawn in the shape of a coloboma radius extending from the pupil at the 7 o'clock position have been used in campaign posterss promoting the search for her.

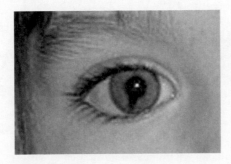

The right eye of Madeleine McCann

Translocations involve pieces of a chromosomes becoming detached from the original chromosome and becoming attached to a different chromosome. This piece of chromosome might disrupt a gene or may contain genes which, when moved to another chromosome, become more or less active leading to different amounts of protein. This occurs, for example, when a specific part of chromosome 8 containing the c-myc gene is transferred (i.e. translocated) to chromosome 14 resulting in a change in the amount of a protein important in controlling cell growth and proliferation and so leading to Burkitt's lymphoma. This is a B-cell lymphoma, first described in 1956 by the British surgeon Dennis Burkitt whilst in equatorial Africa.

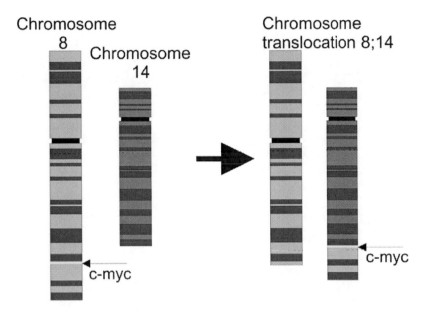

In Burkitt lymphoma a piece of chromosome 8, containing the c-myc gene, is transferred to chromosome 8.

Another aberration is seen when a piece of a chromosome becomes inverted. If no genes are affected these may result in no medical problems. Indeed, such chromosome inversions are common differences between the chromosomes of humans and apes and are suspected to have played an important role in speciation since only those individuals with the same inversions can produce normal offspring.

Chapter 2

Patterns of Inheritance

The information encoded in your DNA determines your unique biological characteristics, such as sex, eye colour, age and Social Security number.

Dave Barry (joke)

The DNA making up our genes are around 99.9% identical among individuals of all nationalities. While most gene sequences are almost identical between all people, some genes can slightly differ among individuals and these are known as alleles (Gr. *allelos*; other). Alleles are forms of the same gene (or any DNA sequence) with small differences in the nucleotide sequence known as either polymorphisms if they are common in a population, or mutations if they are rare. A population of organisms typically includes multiple alleles among various individuals.

Genetic Fingerprinting

Genetic fingerprinting is a technique to allow one to distinguish between individuals using their DNA. Although humans show identical sequences in the vast majority of the genome some specific sequences can be highly variable. It is therefore the likelihood that unrelated individuals would have different sequences at these particular sites in the genome which is the basis for genetic fingerprinting.

An example of such variable sequences are repeats consisting of between 3 and 5 basepairs. When analysing multiple regions containing variable repeats between two individuals a very high statistical power can be generated as one region containing a certain number of repeats does not relate to the number of repeats in any other region. Therefore, if one region in a certain individual has 15 repeats, a second region 30 and a third region 5, the chance of an unrelated person containing the same numbers of repeats is 1 in several million.

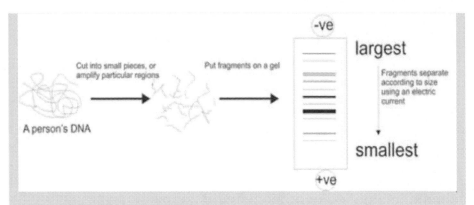

DNA fingerprinting A process by which an individual can be uniquely identified by testing for multiple DNA polymorphisms.

This technique can therefore also allow one to determine if two individuals are related. Naturally, identical twins contain identical genetic fingerprints, while relatives will share varying amounts depending on the closeness of the relationship - For this reason, genetic fingerprinting is often used in paternity tests.

Genetic fingerprinting is also employed in criminal cases to prove that the DNA found at a crime scene is statistically likely to be the same DNA present in a suspect. In 1988, Colin Pitchfork became the first murderer in the UK to be convicted as a result of DNA fingerprinting. Though he had originally evaded police by persuading a friend to give a fraudulent blood sample under his name, he was eventually caught out and sentenced to life in prison for the murder of two young girls. However, controversies do arise from this method, especially when a prosecutor gives some astronomical statistical figure for a DNA sample, which may have been contaminated or mislabelled, belonging to an accused. Indeed, there have been some wrongly ascribed "positive" matches for people who could not possibly have been involved in a crime and it appears several people in the US have been executed on DNA evidence now admitted to be faulty due to human errors in a laboratory. In December 2005 Robert Clark became the 164th person in the USA to be pardoned following post-conviction DNA testing, after serving 24 years of his sentence and in March 2009, Sean Hodgson was released after spending 27 years in jail for murder when tests prove DNA from the scene was not his.

Together, these small DNA differences contribute to each person's unique physical and behavioural features, usually by affecting differences in the amount of a protein produced or resulting in an altered protein which does not function as it should do. When a particular gene is grossly affected by a DNA alteration that has a negative impact on the health of an individual, a genetic disease can present. More than 4,000 genetic diseases have been so far been identified. While most of these are very rare, some are relatively widespread, especially within certain ethnic groups. Around 3% of all children born contain a significant genetic disorder or birth defect and, for individuals under age 25 years old, around 5% develop a serious disease with an important genetic component.

It is important to appreciate that not all mutations result in severe deformities incompatible with life. Mutations serve as the driving force of evolution – as our environment constantly changes, so are we constantly evolving to become more adapted to the environment ("fitter" in the Darwinian sense of the word). For instance there are now several studies linking gene variants to resistance to HIV infection. Not everyone exposed to HIV becomes infected, and while some that are infected become ill relatively soon after infection, for others the onset of ill health can be delayed by up to twenty years. One gene encodes a protein found on the cell-surface of white blood cells which helps to direct the cells to sites of inflammation. The HIV virus uses this protein to enter these white blood cells and an individual with a mutated copy of this gene produces a protein that can deny HIV its route of infection and so slowing down the progression of AIDS. This gene variant is widely dispersed throughout Northern Europe with studies suggesting that it became relatively common about 700 years ago implying that a strongly selective historic event such as an epidemic (there is mixed evidence that this might have been the "Black Death" or smallpox) drove its frequency upwards in populations whose ancestors survived.

In the 1860s an Austrian Monk, by the name of Gregor Mendel, first laid down the laws of genetic inheritance. He studied several characteristics in pea plants such as flower colour. He found that if pea plants producing purple flowers were pollinated by white flowered pea plants, the resulting seeds all produced plants with purple coloured flowers. This result was a complete surprise. Based on ideas about cross breeding at the time the expected outcome would have been either light purple flowers (from a blending of white and purple) or, or a mixture of white and purple flowers.

Gregor Mendel, circa 1865.

Next, Mendel crossed the offspring of this purple-white flower cross (all of which had purple flowers) with each other. This resulted in a population (known as an F2 generation) of plants in which three-quarters produced purple flowers and one-quarter produced white flowers, i.e. a 3:1 ratio.

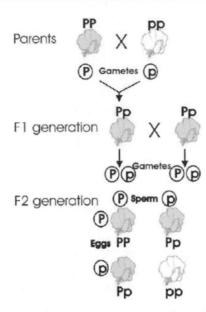

Inheritance of flower colour in the pea plant.

From this data Mendel developed his laws of inheritance, particularly the Law of dominance; when an organism has two different variants (we now use the term alleles) for a trait, the allele that is expressed, overshadowing the expression of the other allele, is said to be dominant (i.e. the purple flower – denoted by P in the diagram). The gene whose expression is overshadowed is said to be recessive (i.e. the white flower – denoted by p in the diagram). Mendel published his findings in 1866 in the little-known journal *Proceedings of the Natural History Society of Brunn*. Even though these were possibly the most exciting findings ever in the field of science, signalling the birth of genetics, they had little impact being cited only three times over the next 35 years. It was not until 1900 that other scientists rediscovered his work and recognized its significance. However, Mendel had long since died in 1884 from chronic nephritis.

How does this inheritance of flower colour in pea plants relate to us? Sexual reproduction in humans occurs with the fusing of two gametes (the oocyte and sperm) each with one half of the pairs of 23 chromosomes. The resulting zygote therefore inherits one copy of a chromosome, with the respective genes, from each parent. When an identical mutated gene or allele is inherited from each of the two parents, the individual is referred to as a homozygote as they have two copies of the same gene; a heterozygote refers to the presence of only one copy of a particular allele. A Mendelian trait or disease is one that is controlled by a single gene mutation and shows a simple Mendelian inheritance pattern – either dominant or recessive. Examples include sickle-cell anaemia, Tay-Sachs disease, cystic fibrosis and xeroderma pigmentosa. In contrast there many disorders that involve a number of different genes (referred to as polygenic) or are influenced by other genes in combination with the environment, which are termed multifactorial and inherited in a non-mendelian fashion.

One example of a dominantly inherited trait can be seen in the presence of the Habsburg jaw in the family tree of the Habsburgs. They passed on a gene for an inherited condition known as mandibular prognathism which, in mild forms, is relatively common and causes a jutting jaw or drooping lower lip. This shows a pattern whereby the gene allele for the

Habsburg jaw (H) is dominant over a normal shaped jaw (h). Therefore, heterozygotes (H/h) containing both an allele for the Habsburg jaw and one for a normal jaw will nevertheless be born with the Habsburg jaw. If two heterozygous (H/h) parents produce children, then there is one chance in four of any of their offspring inheriting both attached alleles (h) and so having an abnormal jaw; the classic 3:1 ratio among offspring of heterozygous parents.

Parent 1: heterozygous (Habsburg jaw)

Parent 2: heterozygous (Habsburg jaw)		H	h
	H	H/h (Habsburg jaw)	H/h (Habsburg jaw)
	h	H/h (Habsburg jaw)	h/h (normal jaw)

If one parent is homozygous for the Habsburg jaw (H/H), even if the other parent has a normal jaw (h/h), then all children will have the Habsburg jaw, i.e. they will be heterozygotes (H/h).

	H	H
h	H/h	H/h
h	H/h	H/h

If both parents have normal jaws, and are therefore homozygotes (h/h), then all offspring will have normal jaws also. However, if one parent is heterozygous (H/h) while the other has a normal jaw (h/h), then there is a 50% chance that any child will have a normal jaw or the Habsburg jaw, a 1:1 ratio.

	H	h
h	H/h	h/h
h	H/h	h/h

The inherited characteristic of the Habsburg jaw or lip had been passed through generations of the family for as far back as 1421 when a jutting jawed princess Zimburg of Massovia married one of the Habsburg princes. For the next 300 years, at least, the jaw deformity would remain a defining feature of one of the greatest European dynasties. Indeed, it appears evident from portraits that members of the family were very keen to parade their underbites as a mark of royalty. If we look at just the Spanish members of the Habsburgs, starting with the aptly named Philip 'the Handsome' and his wife Joanna 'the Mad' of Castile, we can see that both male and female members of the family at each generation were affected, ending with the unfortunate Charles II. Charles II's parents, who were uncle and niece, both appeared to have the Habsburg jaw and so possibly passed on a double dose of the gene to their hapless son who was consequently born with a hugely misshapen jaw that prevented him from eating or speaking properly. As a consequence of his deformed head and the familial madness from Joanna he was unable to produce an heir to the Spanish throne.

Instead, throughout his rule he became increasingly mentally and physically decrepit and his eventual death precipitated the War of the Spanish succession which would last for over a decade and involve much of Europe.

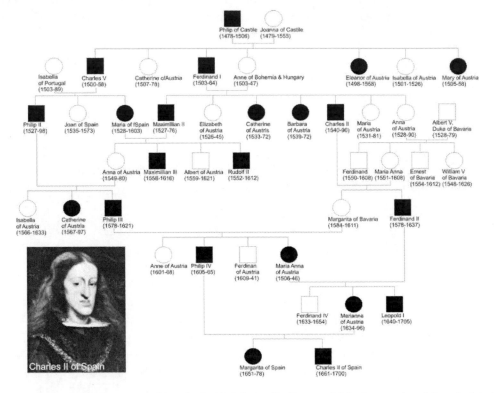

Inheritence of mandibular prognathism in the Habsburg family tree. Male (squares) and female (circles) members appearing to show the trait in portraits are shaded.

HERITABLE FACIAL VARIATIONS

In the same way as the Hapsburg jaw, there are a number of other facial characteristics that are inherited due to the presence of a single gene in a dominant or recessive pattern, i.e. Mendelian inheritance. One example can be the appearance of earlobes; some people have the lobe hanging free (detached ear lobes) while other have the lobe attached to their head. This tends to show a pattern whereby the gene allele for detached earlobes (E) is dominant over attached (e). Therefore, heterozygotes containing both an allele for detached and one for attached earlobes have detached ears. And if two such heterozygous parents produce children, then there is one in four chance of any of their offspring inheriting both attached alleles and so showing attached earlobes; the classic 3:1 ratio. Such an inheritance for the different earlobes can be seen in many families. One example is the British royal family. Judging from photographs, Prince Andrew and Sarah Ferguson both have attached ear lobes (suggesting that they are e/e), and their two daughters seem to have inherited the same trait. This would imply that both Queen Elizabeth (her sister had attached ear lobes) and the Duke of Edinburgh are heterozygotes (E/e), as one of their four children (Andrew) shows attached ear lobes.

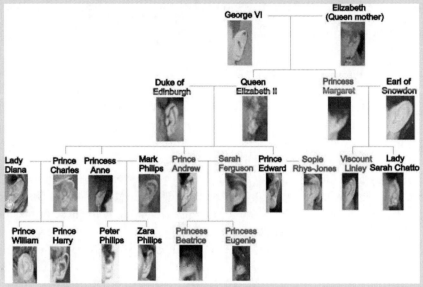

Inheritance of earlobes in the British royal family. Names in red denote attached ears.

There are a number of other, easy to notice, characteristics that are inherited in a similar way. One example is the hairline of an individual that is either straight or else forms a widow's peak, in which it comes to a point in the centre of the forehead. This is thought to be due to a single gene, where a widow's peak results from inheriting the dominant allele, in contrast to a straight hairline which is caused by inheriting two recessive alleles. The occurrence of freckles also appears to be due to a single gene, where having freckles is dominant and not having freckles is recessive; though numbers of freckles can be affected by the environment, such as sun exposure. Another facial characteristic is the cleft chin, which is dominant over a smooth chin. However, not everyone with the gene will show the character, a phenomenon known as 'variable penetrance'. An example is Kirk Douglas and his son Michael, who both have a cleft chin. Having dimples in your cheeks when you smile is also a dominant trait, though, this again exhibits variable penetrance.

Recessively inherited disorders only result if two copies of the defective gene are inherited. Inheriting only one defective copy will not lead to any ill-effects as a second functional copy of the gene will compensate. So while a dominant disease requires only one parent to carry the gene, a recessive disease can only result if both parents have at least one copy of the defective gene. Furthermore, only 25% of the offspring of two such heterozygous (i.e. they only carry one mutant copy of the gene) parents will show the disease. We can see this if we assume the dominant non-disease causing gene as the 'D' and the recessive, disease-causing, gene as 'd'.

	Parent 1: heterozygous	
	D	d
Parent 2: D	D/D	D/d
heterozygous d	D/d	d/d (disease)

   In recessively inherited disorders the parents are often unwitting heterozygous carriers of a gene in which children inheriting two copies suffer from the disease. The family tree of a recessive gene is very different from a dominant one, as only one or two generations will suddenly show the disease when individuals with the same gene marry. This is illustrated by a Mexican family tree containing a gene for the bone disorder Pycnodysostosis (described in *Skeletal Disorders*). This shows that inbreeding between the family members increased the cases of children born with the disorder. The French painter Henri de Toulose-Lautrec was born from a first-cousin marriage with this recessively inherited disease. His parents must have both had a copy of the mutated gene for the disease and at least three of his cousins, the offspring of his paternal uncle married to his maternal aunt, also suffered from the same disorder, supporting the mode of inheritance being autosomal recessive and exemplifying the risks of inbreeding.

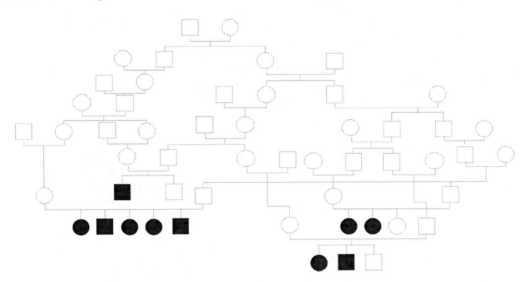

An example of recessive inheritance. Shaded squares and circles denote males and females respectively, who are affected by the recessive disease.

Inbreeding is how we get championship horses.

Carl Gunter, (Louisiana state representative, explaining why he was fighting a proposed antiabortion bill that allowed abortion in cases of incest)

   The ancient taboo of incest in most cultures probably stems from the realisation that inbreeding, particularly within the first generation, leads to higher occurrences of children born with defects. The reason inbreeding is so deadly to a population is that, while normally only a small minority of the general population will carry a copy of a particular gene causing a specific recessive disease (i.e the chance of marrying a carrier of the same recessive gene is low), this chance will increase if the individuals are related and hence share some genetic material. Therefore, some recessively inherited conditions are found in higher frequencies in some close-knit communities such as Amish or various Jewish communities. Another example of inbreeding leading to higher incidents of genetic diseases can be seen in the Negev Bedouin people, a population of 140,000 people who roam the Negev desert in Southern Israel. Their tradition is to marry within the family so as to strengthen bonds among extended families; around 65% marry their 1st or 2nd cousins. Although they do not carry any more mutations than the general population, because so many marry relatives they have a higher chance of marrying someone carrying the same mutation increasing the odds of producing a child with a recessive genetic disease. In fact, many of the diseases seen in the Bedouin are extremely rare, resulting from mutations not previously seen in other populations of the world, due to their cultural isolation. Charles Darwin, along with his cousin Francis Galton who coined the term eugenics, was one of the first to realise the ill effects of inbreeding and the advantages of cross-breeding. His theories may have actually stemmed, to some degree, from his own experiences as he himself married a cousin Emma Wedgwood in 1839. They had ten children with seven either dying prematurely or remaining childless; although dozens of his descendants nevertheless became eminent, particularly in the fields of science and medicine.

When a gene for a genetic disease occurs on the X chromosome as opposed to one of the autosomes, X-linked recessive inheritance is seen (X-linked dominant inheritance is rare). This inheritance is characterised by a high preponderance of male sufferers. This is in contrast to the inheritance of traits on autosomal chromosomes, where both sexes have the same probability of expressing the trait. The reason females are less affected by X-linked recessive diseases is due to the presence of two X chromosomes. They can carry the affected gene on one X chromosome without developing the disease but can pass it on to their sons, i.e. women are unaffected heterozygotes, or carriers. The British royal family again provides an example with the inheritance of the X-linked recessive disease Haemophilia. Queen Victoria was almost certainly a heterozygous carrier of the haemophilia gene, which she passed on to one of her sons, Leopold, who subsequently suffered and died from the disease. Interestingly, Queen Victoria had no ancestors with the condition and so serves as an example of how a disease can arise due to a spontaneous mutation; although there are suggestions that she may have been illegitimate. Prince Leopold then passed on his mother's gene to his daughter, Alice, who subsequently gave birth to a haemophilic son. Two more of Queen Victoria's daughters, Alice and Beatrice, were also carriers of the haemophilia gene, which they transmitted to several royal families in Europe, including Spain and, more famously, Russia with the birth of Alexis, the son of Tsar Nicholas II of Russia. Desperate to cure their son the Russian royal family turned to the 'mad monk' Rasputin whose influence over the family is often cited as a main catalyst for the Russian revolution. His treatment involved hypnosis, a calming confident presence around the boy and family, and exclusion of all the other court physicians who may have been stressful for the young Alexia. It is now known that hypnosis and reduction of stress narrows minor arteries slowing blood circulation and so relieving symptoms. Although it seems unlikely that Rasputin was aware of this, he was certainly not slow to capitalise on his good fortune by spreading his influence over the royal household.

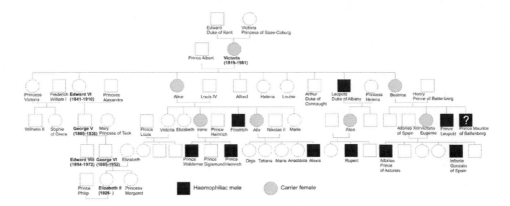

The inheritance of haemophilia in European royal families.

In addition to the haemophilia gene, Queen Victoria may have also carried the gene for another blood disorder called porphyria (described later) supporting her worries regarding the quality of the blood of the British royal family. Perhaps she realised, as did Darwin, the dangers of inbreeding; in a letter to one of her children she describes the necessity of revitalizing what she called the "*lymphatic*" blood of their houses.

"I do wish one could find some more black eyed Princes and Princesses for our children! I can't help thinking what dear Papa said -- that it was in fact when there was some little imperfection in the pure Royal descent that some fresh blood was infused... For that constant fair hair and blue eyes makes the blood so lymphatic."

Mitochondria are organelles that inhabit each cell of our body and function to provide energy. Appearing as numerous small bodies, resembling threadlike grains (Gr. *mitos*; thread, *khondrion*; granule) they carry their own genes and DNA (mtDNA) and divide independently of the cell, supporting the idea that they were originally derived from bacteria. During sexual reproduction mitochondria are passed on to offspring only through the egg and not the sperm as it is only the nucleus of the sperm that enters the egg during fertilisation. Therefore, as only females can pass on mitochondria, the mtDNA we carry in our cells has been carried through a direct line from our very distant maternal ancestors. This is the basis of the 'Eve Theory' in which all mitochondria present in the whole human population originate from a single, or small handful of, women who lived around 100,000 years ago.

When mutations occur in mtDNA these changes are passed on only through females. This can allow one to determine to what extent individuals share the same maternal ancestor. When Oprah Winfrey had her mitochondrial DNA tested in 2005, she was pleased to discover that she was a descendant of the Zulu tribe in South Africa. Perhaps a more interesting example is the testing of the late Russian imperial family. One of the Tsar's daughters, Anastasia, was presumed executed along with the rest of her family on July 17, 1918. However, rumours spread that Anastasia might have survived and in 1991, when the bodies of the Russian royal family were exhumed, Anastasia's bones were indeed missing. In 1922 a woman, later called Anna Anderson, claimed to be Anastasia, and although she could not speak a word of Russian, many people were convinced. To determine if her claims were true her mtDNA was taken after her death, and compared to mtDNA extracted from the bones of Anastasia's mother, the Czarina. As a control, mtDNA from England's Prince Philip, who is related on the maternal side to the Czarina, was compared to the Czarina and indeed showed significant relatedness. However, Anna Anderson's mtDNA showed very little similarities to the Czarina's mtDNA leading one to conclude that she was not Anastasia. One possible identity of Anna Anderson was a woman by the name of Franziska Schanzkowska, a Polish factory worker who disappeared at about the same time that Ms. Anderson appeared in Germany claiming to be Anastasia. Analysis of mtDNA from a grandnephew of Franziska Schanzkowska revealed strong similarities, suggesting that Anna and Franziska were the same person.

Anastasia Nikolaevna Romanova, 1910.

Portrait of Thomas Jefferson. Rembrandt Peale, 1800.

In a similar way to mtDNA, the Y chromosome is passed intact from father to son and so can be used to trace paternal lineages. As the Y chromosome also acquires mutations through time, these sequence differences can again serve as 'markers' varying among individuals and allowing Y chromosomes to be distinguished. The identification of Thomas Jefferson, the 3rd president of the USA, as the father of at least one son of his slave, Sally Hemings, was performed using Y linked markers. As Jefferson had no sons by his wife, Y chromosomes were studied from descendents of his brothers which when compared to DNA from the offspring of Eston Hemings, his presumed illegitimate son, showed significant similarities.

There are a set of specific diseases that show the phenomenon of anticipation whereby the symptoms of a genetic disorder become apparent at a progressively earlier age as it is passed on to each successive generation. This is common in trinucleotide repeat disorders where a dynamic mutation in the DNA occurs known as a triplet repeat expansion mutation. Such disorders include Huntington's disease, myotonic dystrophy, Friedreich's ataxia, fragile X-syndrome and some spinal cerebellar ataxias. Triplet repeat expansion mutations are unusual genetic changes whereby a segment of DNA containing a certain number of a repeat of 3 nucleotides (triplet repeat), such as CAGCAGCAG, increases in length as it is passed from parent to offspring. This type of mutation underlies a number of disorders whereby healthy individuals have a variable number of triplet repeats, but there is a threshold beyond which a high number of repeats cause disease. In this way, the condition may worsen or have an earlier onset from generation to generation. One of the most common diseases is Fragile X syndrome, which develops due to an expansion of the triplet CGG (CGGCGGCGGCGG, etc.) disrupting a gene on the X chromosome and leading to mental retardation. Huntington's disease (HD) is also caused by a repeated triplet (CAG) occurring in the Huntington gene, on chromosome 4. This results in an altered Huntington protein which clumps together inside neuron cells leading to death. This particularly affects neurons in brain areas responsible for mental abilities and movement coordination with the progressive neuronal loss leading to the gradual development of abnormal movements and changes in cognition, behaviour, and personality. A persons' ability to walk, think, talk and reason slowly diminish. The onset of symptoms is usually between the ages of 30 and 50 and it affects between 1 and 2 in 20,000 individuals.

HD, sometimes known as Huntington's Chorea, has been reported since the 16th century and it was the early Renaissance physician Paracelsus who used the term "*chorea*" (the Greek word for dance) to describe the shaking and twitching that people with the disease went through. English colonists in America called the disease Saint Vitus Dance (St Vitus is the patron saint of dancers), and many of the "witches" at the infamous Salem Witch Trials of 1692 in Massachusetts, are now believed to have had HD; their choreic movements and odd behaviour were seen as possession by the devil. It was in New York, in 1872, that George Huntington wrote a paper on the chorea emphasizing the hereditary nature of the disorder and providing the first scientific description of the disease. He was only 22 years old at the time! The family he studied were ancestors of a man by the name of Jeffrey Francis who emigrated from England, carrying the Huntington mutated gene, in 1634. This is an example of a "founder effect" (described in more detail later) which also occurred in South Africa where several hundred people who developed HD were all descendents of a Dutch immigrant called Elsje Cloetens, who arrived there in 1652. Another population with a high incidence of HD live near Lake Maracaibo in Venezuela and are all related to a Spanish sailor from Hamburg, Antonio Justo Doria, who lived during the 18th century and who travelled to Venezuela to buy dye for a German factory. It was the analysis of shared DNA sequences between these Venezuelan descendents that enabled scientists to 'map' the location of the Huntington gene to chromosome 4, by using a process known as Linkage analysis (refer to *Gene Mapping* below).

Woodie Gunthrie, 1943.

In America HD is referred to as "Woody Guthrie's disease" from the famous American folk singer who died from it. One day in the 1950s Woody Guthrie's wife noticed her husband walking lopsidedly. This was soon followed by slurred speech and uncontrollable rages. Eventually, he lost all ability to talk, read, or walk and the only way he could communicate with his wife and children was by waving his arm at cards printed with the words '*Yes*' and *No*'. In 1967 he died from the disease, which had previously killed his mother, from whom he had inherited it. This is because HD is inherited in an autosomal dominant fashion and so a parent with the HD gene has a 50% chance of passing it on to each offspring. When a parent develops symptoms of HD their offspring have to decide whether or not to have themselves tested for the presence of the mutant gene. Should the results prove positive they would live with the certainty of developing symptoms of the disease before the age of 50. Insurance companies now legally have the right to know the outcome of any such test.

MAPPING THE GENOME

Gene mapping involves the idea of fixing a gene to its correct location on one of the chromosomes. This is important to allow further studies of the gene, its inheritance, and possible diagnostic techniques if the gene in question, when mutated, leads to a disease.

In 1990, the U.S. Human Genome Project began and ended in 2003 with a draft version of the sequences of the 3 billion chemical base pairs making up human DNA. This map of the human genome has actually been under construction for the last 100 years beginning with the mapping of the first gene in 1911. This was the gene responsible for red-green colour blindness, which was assigned to the X chromosome following the observation that this disorder was passed on in an X-linked inheritance pattern to sons by unaffected mothers..Subsequently many other disorders affecting only males were likewise mapped to the X chromosome.

The other 22 pairs of chromosomes remained virtually uncharted until the late 1960s, with the discovery of somatic cell hybrids. This involved fusing together human and mouse cells to create unstable cells which quickly lost most of the human chromosomes until only a few remained. Any human proteins in these hybrid cells thus had to be produced by genes located on one of the remaining human chromosomes. This strategy has led to the assignment of around 100 genes to specific chromosomes.

In the 1970's came the discovery of a staining technique called G-banding, from the Giemsa dye, which made the identification of each chromosome much easier with characteristic darker and lighter bands providing the equivalent of latitudes and serving as rough landmarks on the chromosomes. This led to the assignment of some 1,000 genes to specific chromosomes such as the gene leading to Tay Sachs disease (described in *Metabolic Disorders*) found on a particular area of chromosome 15.

A further development of this technique relied on known variable sequences in the genome sequence as markers for nearby abnormal genes. This was first developed by Kan and Dozy, in 1978, who noted that near the sickle cell gene was a stretch of DNA which varied (i.e. polymorphic) between most Africans and African-Americans and which could be used as a marker in a given family for linkage with the sickle cell gene. This polymorphism served as the first use of these common molecular differences between one person and another as a method for both diagnosis (by being linked to a disease gene) and subsequently for general mapping over the entire genome i.e. finding polymorphic sequences of DNA with no known function (a marker) and linking them with the inheritance of a disease in a family. The identification of these markers was helped by the use of special enzymes found in bacteria, known as restriction endonucleases, which cut DNA at a specific sequence. This technique could be used to analyse very large families to determine which individuals inherited a disease in combination with one of the many different markers on different chromosomes. As such, an explosion in the knowledge of genes' chromosomal whereabouts occurred with a little less than 2,000 genes being mapped by 1991. This included pinpointing the location of the Huntington's disease gene (Chromosome 4) using the families in Venezuela, adult polycystic renal disease (Chromosome 16) and cystic fibrosis (Chromosome 7) which were all mapped using this technology.

Meanwhile, scientists learned to sequence these genes from the mid-1970s when Frederick Sanger at Cambridge University and Walter Gilbert and Allan Maxam at Harvard University developed methods for determining the order of bases in a strand of DNA. Automated high-speed sequencing by machine followed in the 1980s, and currently new technology promises to allow the sequencing of an entire genome in a matter of weeks.

It is not the strongest of the species that survives, nor the most intelligent that survives. It is the one that is the most adaptable to change.

Charles Darwin

In 1634, a woman by the name of Anne Hutchinson stepped off a ship from Alford, England to join the other early settlers of the Massachusetts Bay colony. While presently regarded as one of the most important figures in American history being an early champion of religious freedom and women's rights, at the time she did not achieve such recognition. Her opposition to some of the puritan ideals on which that colony was founded lead to her eventual trial. Around this time she delivered an abnormal conception, described by John Winthrop, governor of the colony, as *"not one, but thirty monstrous births or thereabouts, at once; some of them bigger some lesser; few or any perfect shape, none of them at all of human shape"*. The consequences of this for Anne Hutchinson was that she was condemned as being an instrument of the devil and banished from the colony, to be shortly after scalped and killed by Indians along with several of her children. Three of Anne's descendents (Franklin D. Roosevelt, George H. W. Bush and George W. Bush) later became presidents of America.

Among all the fanciful speculations at the time there was a remarkable insight by the local minister, Reverend John Cotton, about the true cause of this abnormal birth. In a sermon he declared that the delivery of the *"thirty monstrous births"*, that we can assume resembled something like a bunch of grapes, was the product of *"several lumps of man's seed, without any alteration or mixture of anything from the woman"*. More than three centuries were to pass before his amazing assertion was actually confirmed. The product of such a pregnancy, known as a hydatiform mole, happens when fertilization occurs in the absence of either a maternal or a paternal set of chromosomes. For instance, this can be seen if an empty egg is fertilized by two sperm - not being viable this forms a mass of cells in the uterus. Such molar pregnancies occur in around 1 in 1,000 pregnancies, and it seems that Anne Hutchinson's was one early description of one.

The phenomenon of molar pregnancies illustrates two points: firstly, two sets of chromosomes are needed at fertilization for an embryo to develop and, secondly, the chromosomes have to come from a male and a female. A foetus of any mammal cannot develop when both sets of chromosomes derive from the same parent. A molecular mechanism has evolved to ensure this does not happen. A small number of particular genes are modified and expressed according to their parent of origin, i.e. they are only functional on either a maternally or a paternally inherited chromosome. This is called genetic imprinting. This means that a foetus inheriting two maternal sets of chromosomes will have twice the normal expression of some imprinted genes that are only expressed on the maternal chromosomes and a complete lack of proteins that would normally be expressed from the paternal chromosome – such a scenario is not compatible with life.

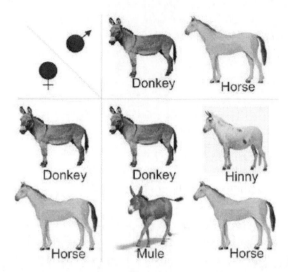

The mating of a female donkey with a horse produces a hinny, whilst a male donkey crossed with a horse gives rise to a mule.

Found only in mammals, this phenomenon of gene imprinting has only relatively recently become understood. However, the idea that maternal and paternal chromosomes differ is an ancient one. For instance, this can be seen in the old practice of mating a female horse with a male donkey to produce a mule; an animal prized for its very hardy temperament. In contrast, mating a male horse with a female donkey produces a hinny. These animals, which like mules are sterile, are generally smaller than a mule and have heads more similar to a horse with smaller ears. This highlights the fact that the genetic contribution from male and female sexes of the same species of an animal is different. George Washington, who liked to breed mules, also realised this. After begging the King of Spain to oppose the ban on exporting the huge Catalonian donkeys, he was eventually rewarded with a Spanish donkey, called the Royal Gift, which he used to sire many mules. Quickly gaining a reputation for being impressively tough, it soon seemed like half the farmers in Virginia owned one of Royal Gift's descendants.

Gerald Campion as *'Billy Bunter'*.

One theory proposes that some genes provided to the offspring by the father try to extract as many resources from the mother as possible while maternally inherited genes, by contrast, are not interested in exhausting maternal resources prematurely, but rather in ensuring that these resources are conserved for future offspring. It is therefore suggested that evolution selected for the phenomenon of genetic imprinting in order to control this. While only a small percentage of our genes are imprinted, mutations affecting one of these will lead to disorders which differ depending on whether the gene defect was inherited from the mother or father. For example, an individual missing a small piece of chromosome 15 from the mother will be born with Angelman syndrome, manifesting itself with mental retardation. In contrast, if the abnormal chromosome 15 is inherited from the father the offspring will present with Prader-Willi syndrome characterised by lower IQ, small stature and a tendency for overeating, leading to obesity, later in life. Such individuals incorrigibly steal and hide food, like the fictional scallywag Billy Bunter who, with his inept ability for snaffling food, displays characteristics of Prader-Willi syndrome. The idea that some genes, behave differently in an offspring if they derive from either the mother or father has opened up new fields of research in genetics concerned with inheritance of characteristics which are independent of DNA sequence. This type of inheritance, the transmission of non-DNA sequence information through meiosis or mitosis, is known as epigenetic inheritance (Gr. *epi*; on/over).

The molecular basis of epigenetics is complex and involves modifications of the activation of certain genes, but not the basic structure of DNA. This can play a role in what has come to be known as cell memory. A genome can pick up and lose epigenetic markers, such as regions of methylated DNA (describing the addition of methyl groups to cytosine residues in the DNA) or remodelling of the chromatin proteins that wrap around the DNA, more rapidly than it can change DNA sequence. This allows cells and organisms the ability to quickly respond to the environment in an inheritable way. This resurrects the idea of the inheritance of acquired traits which was once widely discredited, partly because it seemed at odds with the much more widely accepted theory of evolution envisaged by Darwin. Inheritance of acquired traits was first proposed by the French naturalist Jean Baptiste de Lamarck who speculated that environmental cues could cause phenotypic changes transmittable to offspring. To explain this he used the evolution of the giraffe as an example. He speculated that giraffes once had necks no longer than those of zebras, but as the early giraffes stretched their necks to feed from the highest limbs of a tree, their necks got longer and longer. This acquired trait was then presumably passed on to subsequent generations of giraffes who would be born with progressively longer necks. Sadly, the use of this seemingly absurd example, invited criticism and even scorn from contemporary scientists. The German scientist August Weissman "tested" Lamarck's theory by cutting off the tails of generations of mice and proclaiming that none were ever born without tails. Interestingly though, it is speculated that his experiment when conducted over many generations of mice might, theoretically, lead to tailless mice as his tail amputation procedure could introduce infections, and so a mouse born without a tail, even if an extremely rare event, might have a greater advantage.

During the Second World War, a German-imposed embargo in an area of Holland, already suffering from severe food shortage and the onset of a very harsh winter, led to the death by starvation of over 30,000 people. Analysis of birth records from this time reveal a link between prenatal exposure to this famine and the development of various health problems in the subsequent children such as low birth weight, diabetes and obesity suggesting that a pregnant mother's diet can affect her children. This could give some credence to popular beliefs relating to birth signs, i.e. the idea that the time of year of birth affects personality. Children born in the winter months would have been exposed to a different prenatal-diet than those born in the summer (though in modern times this effect is less pronounced since the food available in supermarkets hardly varies seasonally). A particularly significant finding of this so-called "Dutch Hunger Winter Study" was that the grandchildren of women who experienced starvation tended to have smaller birth weights supporting the idea of the inheritance of characteristics acquired during lifetime.

Portrait of Jean-Baptiste de Monet Chevalier de Lamarck. Charles Thevenin, 1803.

All this came too late for Lamarck who sadly died penniless in 1829 and was buried in a rented grave. However, he is now starting to receive the recognition he deserved for his contribution to natural science.

Chapter 3

SKELETAL DISORDERS

The evil that men do lives after them; the good is oft interred with their bones.

William Shakespeare. Act 3, scene ii, Julius Caesar.

Bone in the body is a living tissue with a vascular and neural supply and performs many functions in addition to providing mechanical support for the body. Bones are composed of a hard matrix of minerals (such as calcium and potassium salts) deposited around protein fibres (such as collagen). In the developing foetus, the skeleton is first laid down as cartilage which is then replaced by bone. Bone X-rays, especially in young children, will still show areas of cartilage present at the ends of growing bones; this can be used to estimate a child's age. However, even in adulthood, the size and shape of bones change, and are continuously being remodelled in response to everyday stresses. In fact, the entire human skeleton is replaced every 10 years, through the activities of bone-dissolving and bone-rebuilding cells. Excess activity of the bone dissolving cells, which can occur in women after menopause, causes osteoporosis (Gr. *osteo*; bone, *poros*; pore) where the bones become weakened. In contrast, reduced activity of these cells can lead to osteopetrosis (Gr. *petros*; stone) resulting from too much bone forming, and leading to extra-dense bone that is more brittle and fractures easily.

Skeletal defects are common birth abnormalities and can be caused by a variety of different genes leading to more than 200 distinct diseases characterised by decreased, increased or deranged bone growth.

Decreased bone or cartilage growth usually leads to disproportionate short-stature, known as osteochondrodysplasias (Gr. *chondro-*; cartilage, *dys;* bad, *plasia;* growth), that differ from proportionate short stature seen in disorders such as those involving growth hormone deficiencies. Depending upon which bones are primarily affected, such as in the limbs, trunk, or ribs, it is possible to differentiate by a glance some of the various types of disproportionate short-stature, and in some cases which gene might be involved.

Jason Acuña, 2009.

Achondroplasia (Gr. *a;* lack of, *chondroplasia*; cartilage growth), for example, is characterised by a normal body trunk but shortened arms and legs. This is because cartilage cells, during development, develop into bone more slowly than normal which becomes especially evident during the formation of the long bones of the arms and legs leading to their reduced length. It is mutations in the fibroblast growth factor receptor gene 3 (FGFR3) that leads to this disorder. Normally this gene produces a protein functioning to limit the formation of bone, however the achondroplasia mutation results in a protein being produced which is overactive, and so signalling even less growth. Achondroplasia is the most common form of inherited disproportionate short stature affecting around 1 in 26,000 people. Interestingly, even though this disorder is inherited, many people with achondroplasia have parents with normal stature. This is because a large proportion of individuals with achondroplasia acquire the disorder as a result of a *de novo* gene mutation occurring when the gene is passed from a parent to the offspring. Such *de novo* gene mutations are associated with advanced paternal age as previously described. While achondroplasia is autosomal dominant – an individual only needs one copy of a mutant gene - inheriting two copies of the mutant gene leads to thanatophoric dysplasia (Gr. *thana*; death, *tophoric*; bringing), which is a lethal condition.

There are many well-known people with achondroplasia such as the actors Josh Ryan Evans and David Rappaport, and Jason 'wee man' Acuña the star of *Jackass* and a professional skateboarder. Achondroplasia was unusually common in ancient Egypt. Numerous mummified bodies and skeletons currently in museum collections show evidence of the condition, and there are also scenes adorning tomb walls depicting people, with short stature. This is thought to result from the fact that Egypt was a closed society for quite some time and there was a total acceptance of achondroplasia, allowing individuals with the condition to procreate without prejudice. It seems that people with short stature may even have been revered in ancient Egypt; the Egyptian gods, Bes and Ptah, are often depicted as having short stature.

Count Guriev. Ingres, 1821.

While achondroplasia results from an overactive FGFR3 protein, mutations leading to a less active version of this protein (and the related FGFR2) signal increased bone growth resulting in the premature fusion of bones, particularly in the skull. Normally, during growth, gaps known as fontanelles exist between the bones allowing the skull to develop. However, if these gaps fuse too early, the skull grows in other directions resulting in an abnormally shaped head, often with bulging wide set eyes, underdeveloped upper jaw, and a beaked nose. Affecting 1 in 60,000 this autosomal disorder is known as craniofacial dysostosis or Crouzon's syndrome. Often surgery is performed to prevent the closure of sutures of the skull from damaging the brain's development and allowing an individual to lead a perfectly normal life. This condition appears to be shown in a painting of Count Guriev by Jean Auguste Ingres in 1821.

There are other types of short-limbed short stature that develop in different ways, due to a variety of different genes. Pseudoachondroplasia (Gr. *pseudo*; false), for instance, was long confused with achrondoplasia (hence the name), though occurs through a very different molecular mechanism. In this syndrome a mutant protein is produced in cartilage cells, which kills a number of them, resulting in slow bone growth.

The Ovitz family, widely known as the *Lilliput Troupe*, had pseudoachondroplasia. These were a successful family of performers who were famous entertainers in Central Europe until the Nazis deported them to Auschwitz in May 1944. Shimshon Eizik Ovitz had the mutant gene which he passed on to seven of his ten children, who were subsequently were born with the condition, as it is autosomal dominantly inherited. Finally liberated by Russian troops, the family eventually found their way to a new home in Israel where they once again became successful performers.

The Ovitz family.

A slightly different type of disproportionate short-stature, also characterised by shortened limbs, is cartilage-hair hypoplasia, so called due to the occurrence of sparse, light-coloured hair. This is autosomal recessively inherited and leads to short stature in addition to sparse, light-coloured hair. The underdevelopment of cartilage in this syndrome, results from mutations in a gene leading to decreased growth by impairing cell division and growth. Two of the most widely known people with this condition are the actors Billy Barty, who founded the organisation and charity *Little People of America*, and Verne Troyer, star of the *Austin Powers* films. Standing only 81cm high, Verne Troyer is currently the smallest actor and actually got his first break in show business as a stunt double for a 9 month old baby in a film called "*Baby's Day Out*".

Billy Barty.

Some forms of disproportionate short stature are characterised by a shortened body trunk, in addition to the shortened arms and legs while the head, hands and feet are of normal size. One type is spondyloepiphyseal dysplasia congenita, a rare autosomal dominantly inherited disorder with which the actor Warwick Davis was born. He is best known as *Professor Flitwick* in the "*Harry Potter*" movies, as the title character in "*Willow*", as Wicket in "*Star Wars*", and as the murderous leprechaun in the horror movies of the same name.

Warwick's height is 106cm and he says of his stature: "*the only real drawback to being small is the associated health problems. As you get older, you can suffer from painful hips, and our joints wear a lot quicker than for people of average height.*" The late American actor Michael Dunn, star of "*The Wild Wild West*" TV series was also affected by this condition which is caused by defects in a gene coding for a collagen protein. Collagen (Gr. *kola*; glue, -*gen*; making, pertaining to the ancient process of boiling animal bones for glue) is the main component of bone and cartilage and is also responsible for providing the strength and elasticity in connective tissues, organ walls, blood vessels and the skin.

Warwick Davis.

Short ribs and a small chest, in addition to shortened long bones of the arms and legs, define a group of disproportionate short statures termed short-rib dysplasias. The most common example is Ellis-van-Crevald syndrome (EVC), which used to be known as six-fingered dwarfism due to the associated occurrence of polydactyly (Gr. *poly*; many, *daktulos*; fingers), the medical term for the presence of more than the normal number of fingers or toes. EVC was first described by Richard Ellis of Edinburgh and Simon van Creveld of Amsterdam. They happened to be sitting opposite each other on a train to a conference in England in the 1930s. After striking up a conversation, they soon realised that they each had a patient with the similar, as then unknown, syndrome sharing characteristics of short stature, short ribs and small chest and the presence of polydactyly. While relatively rare in the UK (affecting around 1 in 60,000 births), this recessive disorder, has a very high incidence of 1 in 200 in the Old Order Amish of Pennsylvania. This is again due to the previously mentioned founder effect, describing the situation when a small part of a population moves to a new location, or when the population is reduced to a small size, resulting in the genes of the 'founders' of the new society becoming disproportionately frequent in the subsequent population. When individuals in the group marry within it, especially to relations, there is a greater likelihood that the recessive genes of the founders will come together in offspring. In this way, recessive diseases, which require two copies of the gene to cause the disease, will show up more frequently than they would if the population married outside the group. The Old Order Amish population stems from a small number of about 200 immigrants, including Samuel King and his wife who came to the area in 1744; one of these two carried a copy of the gene for Ellis-van-Crevald syndrome.

An Amish child with Ellis van Creveld syndrome.

Polydactyly is not restricted to the aforementioned syndrome and is actually relatively common occurring as non-syndromic forms, i.e. in the absence of any medical defects, in around 1 in 500 people. Often this is seen as familial polydactyly where extra fingers are passed along from parents to their children. It is common in the African American population, typically occurring as a 6th little pinkie finger. Again, restricted breeding in some isolated populations have led to higher incidences. One extreme example was seen in the inhabitants of the village of Eycaux, occupying an inaccessible and mountainous region of France, who at the end of the last century nearly all presented with extra digits.

Antonio Alfonseca.

Nowadays, these extra digits are often surgically removed during early life, such as in the case of the cricketer, Sir Garry Sobers. In previous times though, such operations were seldom performed due to risks of infection; Anne Boleyn supposedly went as far as to change the fashion for extra long sleeves in order to hide her extra finger. Unfortunately however, the extra little finger played a part in witchcraft accusations against her during her trial and subsequent execution. Legend has it that the renowned Italian composer and violinist Giuseppe Tartini had six fingers on his left hand that enabled him to play his famously complicated pieces, such as the *Devil's Trill Sonata*. Though, as attractive as it might sound to a musician, extra fingers have certainly not been appreciated by all. Indeed, the American Blues guitarist and singer, Hound dog Taylor, finally cut off his extra sixth digit in a bar one night, shortly before he died, complaining that it had always hindered his guitar playing, getting caught between strings. Conversely, Django Reinhardt, arguably the greatest jazz guitarist of all time, managed to get away with the use of only two fingers on his chord hand after an accident removed the others. This led to his characteristic style which still has a major influence on many jazz guitarists playing today, including Tony Iommi of *Black Sabbath*. He lost his middle and ring fingers of his right hand at the age of 17 whilst at work in a sheet metal factory. His boss came to the hospital with a Django Reinhardt record for him to listen to which inspired him into becoming one of the most influential guitarists in rock music. *The Grateful Dead* guitarist, Jerry Garcia, also played missing a finger as a result of a childhood accident with an axe. One profession that has attracted people with abnormal numbers of fingers is baseball pitching; supposedly, such individuals can achieve extra and unusual forms of spin on a ball. The current major league pitcher with six fingers on each hand is Antonio Alfonseca who, nicknamed *The Octopus*, also has six toes on each foot - a trait inherited though several generations of his family.

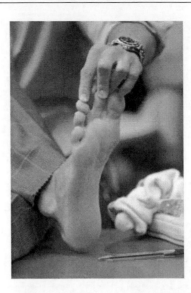

Ashton Kutcher shows his syndactyly on live television.

Polydactyly is not confined to humans, and is often found throughout the mammalian kingdom. Caesar used to ride a polydactyl horse which had feet that were almost human, the hoofs being cleft like toes. As soon as it was born in Caesar's stables, the soothsayers declared that it showed its owner would be lord of the world. Caesar reared it with great care and made sure he was the only person to ever ride it. Another famous polydactyl horse was Bucephalus, the mount of Alexander the Great. Earnest Hemingway famously had a cat with six toed paws. The ancestors of his original cat, many of which have inherited this trait, still roam his old house on Key West which is now a museum.

A related birth defect is the presence of webbed fingers or toes known as syndactyly (Gr. *Syn*; together), in which the digits appear to be stuck together. It results from a failure of programmed cell death that normally occurs between digits during foetal development. In some eastern cultures this anomaly has been associated with divinity and wisdom; Buddha is said to have possessed webbed hands. In western cultures such anomalies tend to be viewed in a more negative way. However, the actions of the celebrities Ashton Kutcher and Rachel Stevens from *S Club 7* fame, who have both separately presented their webbed toes on live television, may have gone some way towards changing people's perceptions. A related disorder characterised by finger webbing and an often disproportional sized hand, affected the late British TV presenter Jeremy Beadle. Known as Poland syndrome, this is thought to result from a temporary alteration in blood flow in the developing shoulder during embryonic development. Another person with this disorder is the American actor Gary Burghoff, best known for playing *Radar* in the "*M*A*S*H*" series.

Ectrodactyly (Gr. *Ectro/ektroma*; abortion, used to explain a congenital absence) is used to describe the absence of one or more fingers or toes. One example of an hereditary form is split-hand/split-foot malformation, also known as lobster-claw deformity, caused by mutations in genes important in controlling growth and patterning of the developing limb. This defect has been documented as far back as 1685 with a strange story set in the remote Scottish village of Wigtown. Two women were executed for religious dissent and, though pleading their innocence, were nonetheless drowned in the Bladnoch River by a particularly cruel executioner by the name of Bell. As one of the women, Margaret Wilson, was being drowned she reputedly called upon the crabs to curse the executioner, and shortly afterwards Bell fathered a child with 'crab hands' - a trait that was subsequently carried down the family line for over 300 years. In the US there is the story of the infamous Grady Stiles, known by his circus as the *Lobster Man*, who had the condition passed on to him through four generations of his family in a dominantly inherited manner. He in turn fathered four children, two of whom also had lobster-hands. While the Stiles family toured for years with a carnival, Grady, who was an alcoholic, used to terrorise and beat his family. This culminated on one occasion when he went as far as to shoot and kill his daughter's fiancé. Somehow he managed to escape legal justice by playing on his deformity though he was not able to avoid the retribution of his wife who persuaded a neighbour, to murder him in 1992. A similar disorder is common among a tribe in Zimbabwe. Early explorers to the region retuned with tales of "*ostrich –footed people*". A number of individuals in this tribe are born missing their three middle toes and again serves as another example of the effects of inbreeding in an isolated population. There are other communities in remote regions of the Kalahari Desert who also share this disorder. It is also possible to have only one effected handor foot. The Soviet-Latvian chess Grandmaster and the eighth World Chess Champion, Mikhail Nekhemievich Tal had ectrodactyly in only his left hand.

Grady Stiles.

Mutations in other genes affecting the developing limb can lead to more severe phenotypes such as acheiropodia (Gr. *a*; absence, *cheiros*; hand, *podus*; foot) characterised by limbs that terminate in a stump. This condition affected Carl Hermann Unthan, known as the *Armless Fiddler*, who was born in Prussia in 1848. From a young age, his father pushed him to do things for himself resulting in his ability to use his feet to grasp things and perform tasks that most people required their hands for, such as writing and even playing the violin - he once played for Strauss. He toured America with an act which included performing card tricks using only his feet. During the First World War he also travelled around war hospitals giving motivational speeches to German amputees. His autobiography, written on a typewriter using his toes, and entitled "*The Pediscript*", was published just after he died at the age of 80. Another sufferer was the French painter, Docournet, who as well as lacking arms, was also born with a crab-like foot, which he used to great effect in holding a paintbrush. One of his most famous pieces, the eleven foot high depiction of Mary Magdalene at the feet of Christ, still resides in his home town of Lille after it was purchased by the government.

Mikhail Nekhemievich Tal, 1960.

Members of the Vadoma tribe with ectrodactyly.

While the aforementioned types of disproportionate short-stature result from decreased bone growth, other forms of osteochondrodysplasias are characterised by decreased bone density. One example is osteogenesis imperfecta (OI) (Gr. *osteo*, *genesis*; formation, *imperfecta*; imperfect), which is commonly known as brittle bone disease or 'glass bone disease'. OI is a group of autosomal dominantly inherited diseases in which collagen formation is disrupted leading to brittle bones. The more severe the defect in the collagen fibres the greater is the severity of OI. While severe forms present with short stature, mild forms can leave individuals with a perfectly normal appearance other than having weak bones.

Henri de Toulouse-Lautrec, circa 1892

Nabil Shaban portraying Ivar the Boneless in the Channel 4 documentary *The Strangest Viking*.

Osteogenesis has been described many times through history. It has been suggested that the Viking chieftain Ivar Ragnarsson, who invaded and became King of England in 865 AD, suffered from one of the more severe types of OI. He was the oldest son of Ragnar Lodbrokby who, according to an ancient Nordic poem from the time, forced non-consensual sex on his new wife, and was punished by having a first-born son, Ivar who was described as being without any bones. Acquiring the nickname Ivar the Boneless, he was unable to walk on his legs and so was carried around on a shield. However, any exact analysis of Ivar's bones, to test if he really did suffer from OI, is out of the question since William the Conqueror is said to have dug up his grave and burnt its contents in a great bonfire to celebrate the Battle of Hastings. Milder forms of OI, in which individuals present with normal height but are vulnerable to bone fractures, when undiagnosed, have led to instances of child abuse allegations. In one of many examples in America, the parents of four-month-old Liliana Valasquez were accused of beating their child when X-rays taken during a check up in 2000 revealed multiple rib fractures. Hospital staff neglected to test whether the child suffered from OI and instead insisted the fractures could only be the result of child abuse. After 18 months in the foster care system, Liliana was eventually returned to her parents, and the criminal charge against the father was dropped. The parents have since had another child with the bone disorder.

In contrast to diseases resulting from decreased bone production, other disorders are characterised by increased production of bone or cartilage and are known as osteopetroses (Gr. *osteo*; bone, *petros*; stone). Also referred to as marble bone disorders, the bones in these disorders are very hard and dense, though also brittle - normal bones are flexible to some degree. These vary widely in their severity. Mild types such as the autosomal recessive Albers-Schönberg disease are relatively common (1 in 20,000), where affected individuals may show no symptoms other that deafness resulting from mild bone overgrowth in the skull compressing nerves. This increased bone growth is caused by reduced activity of bone dissolving cells from mutations in a gene important for their function. A more severe form is pycnodysostosis (Gr. *pyknos*; dense, *dysostosis*; abnormal bone formation), a rare autosomal recessive disorder where, in addition to abnormally dense brittle bone, there is disproportionate short stature. There have been at least 100 cases listed worldwide and parental consanguinity (inbreeding) has been noted in a large number of cases. These patients have mutations in a gene coding for a protein which is normally responsible for the degradation of collagen and other bone proteins. The artist Tolouse Lautrec inherited this disorder. Standing was only 1.53 meters tall with short legs, he suffered numerous fractures associated with bone fragility. However, it was his short stature which resulted in his unique artistic style, particularly the "cut-off" technique subsequently adopted by many other painters, where objects and figures are truncated by the edge of the frame. Sadly, his inherited disorder also inflicted psychological scars leading him to self-destruct and die at the age of 36 from alcohol abuse.

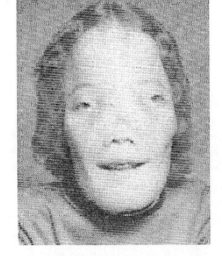

Roy "Rocky" Dennis, 1977.

Craniotubular dysplasias are diseases characterised by abnormal bone shape. These generally result from excessive deposition of calcium causing new bone to form. Mild forms of these diseases, caused by accumulation of calcium in the synovial fluid around the joints, may only result in moderate knee problems. However, the inheritance of a gene leading to calcium deposits building up in the skull, lead to the severe craniodiaphyseal dysplasia (*diaphysis*; the hollow tube of cortical bone), presenting with cranial thickening and the narrowing of cranial foramina (perforations in bones for nerves to pass through) resulting in blindness, deafness and facial palsy. This disorder is depicted in the 1985 biographical movie called *The Mask* relating the life and early death of Roy L. "*Rocky*" Dennis, who suffered from this disease and died in 1980 at the age of 16.

These things are good:
ice cream and cake, a ride on a Harley,
seeing monkeys in the trees,
the rain on my tongue,
and the sun shining on my face.
These things are a drag:
dust in my hair,
holes in my shoes,
no money in my pocket,
and the sun, shining on my face.

Roy "Rocky" Dennis 1964-1980

Another example of abnormal bone growth is sclerosteosis (Gr. *sclero*; hardened, *-osis*; abnormal condition). This autosomal recessive condition is extremely rare though is most prevalent in the Afrikaner population of South Africa as a result of the founder effect. It can be traced back to an individual arriving in the country around the 17th century. Overgrowth and sclerosis of the skeleton, particularly of the skull, occurs in early childhood, and results from a mutation in a gene encoding the protein sclerostin. This protein functions to inhibit bone production by controlling the activity of a group of proteins known as Bone Morphogenic Proteins (BMPs). BMPs play important roles in stimulating the formation of bone during development and in response to injury. A mutation in the sclerostin producing gene causes the body to respond to injuries such as bruises and sprains by growing new bone at the site of the injury rather than the appropriate tissues. Eventually a second skeleton begins to form severely restricting mobility. This disorder is called fibrodysplasia ossificans progressiva (*fibro*; fibrous connective tissue, *ossification*; process of bone formation, progressive) where connective tissues, such as cartilage or muscle are turned to bone. The condition was first reported in the 17th century by Patin, a French physician, who described a condition in which a woman "*turned into wood*" - the wood being new bone deposits. Occurring at a frequency of 1 in 2 million people the most well known sufferer was the American Harry Raymond Eastlack. Symptoms began when he was 10 years old. He started to develop painful nodules in the neck and shoulders - removal of these nodules only caused larger ones to form. This progressed until towards the end of his life he was able only to move his lips. Before he died of pneumonia he requested that his body be donated to science, and his skeleton is still on display at the Mutter Museum in Philadelphia showing how the muscles of his back had turned into sheets of bone.

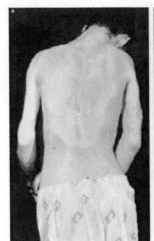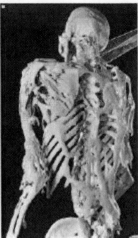

Raymond Eastlack.

CLONING

A clone is a single cell (such as bacteria, lymphocytes, etc.) or a multi-cellular organism that is genetically identical to another living organism. Natural clones occur when an organism reproduces asexually or when two genetically identical individuals are produced by accident. The term *clone* is derived from the Greek word for "twig" which was originally used to describe all descendants of a single plant, produced by vegetative reproduction. In fact many horticultural varieties of plants are clones, with some varieties of grapes representing clones that have been propagated for over 2000 years!

Cloning in the molecular biological sense can refer to a number of different techniques:

Molecular cloning refers to the procedure where by a piece of DNA is inserted into cells which can then be cultured. This allows the inserted sequence to be analysed, such as testing what protein is derived from any genes which might have been in the region.

Cellular cloning describes the creation of a new organism with the same genetic information as a cell from an existing one. This often refers to the process known as somatic cell nuclear transfer in which a cell of the organism to be cloned, with its nucleus containing the DNA, is transferred into an egg cell which has had its nucleus removed. This results in the host egg cell developing into an organism with genetically identical DNA to the donor.

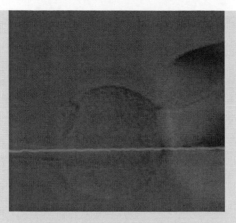

A narrow tube removes the nucleus containing the maternal chromosomes from the oocyte prior to the somatic cell nuclear transfer.

The first animal clone was a frog, cloned by Thomas J. King and Robert W. Briggs in 1952 using somatic cell nuclear transfer. A carp was also created in this way, by inserting DNA of a male carp into the egg of a female carp, in 1963 by an Asian scientist Tong Dizhou, 33 years before *Dolly* the Sheep. Sadly, like Mendel, he published his findings in an obscure science journal which that was not translated into English. Since then there has been *Dolly* the sheep (she was cloned from mammary cells hence the name in honour of the country and western singer, Dolly Parton) in 1996 followed by two more sheep, *Molly* and *Polly* who were the first transgenic mammals containing a human gene. In 1997 came the arrival of 3 mules by the names of *Idaho Gem, Utah Pioneer, Idaho Star* (Americans, naturally!), followed by *Dewey* the deer, *Prometea* and *Paris Texas* the horses, 5 Scottish piglets (*Millie, Christa, Alexis, Carrel, and Dotcom*), an ox named *Noah*, two Jersey cows named *Millie* and *Emma* (there have been several other since), *Little Nicky* and *Copycat* the cats, *Cumulina* the mouse, a rat called *Ralph*, a rabbit which was sadly nameless (developed by the French), a dog named *Snuppy*, a monkey called *Tetra* and, supposedly, a human called *Eva*.

It was in December 2002 that Clonaid, the medical arm of a cult called Raëlism who believe that aliens introduced human life on Earth, claimed to have successfully cloned the first human being, a baby named Evá. They maintained that aliens taught them how to perform cloning, even though the company had no record of having successfully cloned any previous animal. The company has failed to provide the DNA which would allow an independent agency to prove that the girl was a clone and their claims are now generally discredited.

In 2005 came the news that a group of scientists led by Hwang Woo-Suk of Seoul National University in Korea had succeeded in creating human embryonic stem cells by cloning. However, later in 2005 this was also exposed as a fraud.

Chapter 4

CONNECTIVE TISSUE DISORDERS

Indecision is the key to flexibility.

A bumper sticker.

Connective tissue describes the structure surrounding and supporting cells found within tissues. This is composed mainly of structural proteins such as collagen and elastin, and more specialised proteins such as fibrillin. Mutations in genes coding for particular types of collagens (different to those types causing osteogenesis imperfecta) result in Ehlers-Danlos syndrome. Characterised by defects involving the skin and joints the disorder affects around 1 in 5,000 individuals, the severity, characteristics and inheritance pattern (usually autosomal dominant with a small minority showing autosomal recessive inheritance) dependant upon which disrupted collagen gene is inherited. The symptoms involve thin and translucent skin which bruises easily, excessively loose joints, from the larger joints of elbows and knees, to the smaller joints of fingers and toes, and defects in connective tissue of the intestines and arteries.

Nicolo Paganini of Genoa, proclaimed by many as the greatest violinist who ever lived, was said have suffered from Ehlers-Danlos syndrome. He had bony hands, thin, almost translucent skin, large feet and a disproportionately long neck. However, his hyperextensible joints allowed him an extraordinary flexibility and many doubt he could have been so proficient without the abnormal dexterity in his fingers. During his performances he would often contort his body into seemingly impossible poses. The disease also led to numerous recurrent bouts of severe abdominal pain, which would trouble him for weeks to months at a time and leave him weak. *"I am not handsome,"* he used to say, *"but when the women hear me play, they come crawling to my feet"*. His only offspring, Archillino, had none of his father's clinical features. Garry "Stretch" Turner, 39, a pub landlord from Lincolnshire, also has EDS and as a result of the stretchy skin associated with the condition holds the record for having the most clothes pegs attached to his face – 153!

Marfan syndrome, named after the French paediatrician Antoine Marfan, who first described it in 1896, is another inherited disorder of the connective tissue. Affecting 1 in 3,000-5,000 people, it is inherited in an autosomal dominant manner so that in most cases there is also an affected parent.

Nicolo Paganini.

   Marfan syndrome was once also believed to be the result of abnormal collagen, similar to EDS, though it is now known to be caused by mutations in a different gene coding for the protein Fibrillin. This protein allows tissues to stretch repeatedly without weakening, and when disrupted can lead to connective tissues that are looser and weaker than normal. Increased bone length also occurs and the most recognisable signs associated with this syndrome are excessively long arms and legs and often long narrow faces. Curvature of the spine can also develop, though the most important complications of Marfan are those affecting the heart and major blood vessels. The lack of Fibrillin in the walls of the aorta (the large blood vessel that carries blood away from the heart) and mitral valves (valves which separate the chambers of the heart) can lead to them being weak and loose resulting in an increased susceptibility to heart problems and sudden death. As this disorder can lead to increased height and lengthened arms, many sufferers become involved in sports such as basketball and volleyball. However, a missed diagnosis of this disorder can prove fatal, especially during physical activity. One high profile case was that of the American Olympic Volleyball player, Flo Hyman. Standing at six feet five inches tall, she was generally considered the best female volleyball player in the world, leading America to a silver medal at the 1984 Olympics in Los Angeles. Only 2 years later, during a game in Japan, she collapsed and died, with a subsequent autopsy revealing that she had Marfan's syndrome. An unsuccessful proposal was made in the 1980s for athletes in the National Basketball Association in the USA to undergo testing for Marfan syndrome, but this matter is still under discussion.

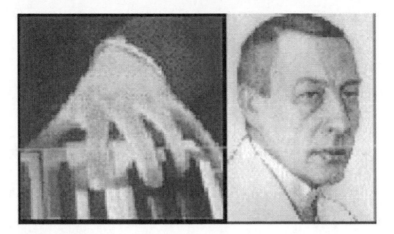

Sergei Rachmaninov.

Marfan syndrome is sometimes called arachnodactyly, (Greek for spider-like fingers) since another characteristic sign is disproportionately long fingers and toes. Sergei Rachmaninov, who is widely suspected to have had Marfan's syndrome, is said to have had one of the widest hand spans of any pianist. He was able to cover a twelfth with his left hand - a span of approximately 12 inches from his little finger to his thumb – well over the length of an A4 page! Marfan's syndrome has also been linked to several other prominent figures, such as Abraham Lincoln, Charles de Gaulle and Osama bin Laden, underlining the importance of height in leadership. Osama bin Laden is said to be around 6 feet 5 inches tall, apparently unusual for his family, with a thin and bony frame and uses a cane possibly as a result of a curved spine. His facial structure, with his elongated head and narrow face is also suggestive of Marfan's syndrome. It is also reported that he sees doctors regularly with heart problems.

Chapter 5

MUSCULAR DISORDERS

The function of muscle is to pull and not to push, except in the case of the genitals and the tongue.

Leonardo da Vinci

Muscle is the contractile tissue found in the body. The two main types are voluntary muscle, also known as skeletal muscle, which is anchored by tendons to the bone and is consciously used to move the body, and involuntarily muscle which moves without any conscious thought, such as in contraction of the heart. Muscle diseases, known as muscle myopathies (Gr. *myo*; muscle), occur when defects in muscle cells lead to skeletal muscles becoming weak or shrunken.

Duchenne and Becker muscular dystrophies (Gr. *dys*; painful/difficult, -trophy; nourishment/ development) are the most common inherited forms sharing similar progressive muscle weakness and wasting symptoms. Both of these disorders are caused by mutations in the same gene, the dystrophin gene on the X chromosome, and are inherited as X-linked recessive disorders that generally affect males. The dystrophin gene is the largest known human gene stretching over 2.3 million bases of DNA - this is almost 0.1% of the entire human genome. This gene codes for the dystrophin protein which helps strengthen muscle fibres, protecting them from injury as they contract and relax. Without this protein muscles become damaged with use causing the characteristic muscle weakness and heart problems seen in Duchenne and Becker. Many different mutations in the dystrophin gene have been identified; the more severe Duchenne mutations typically prevent any functional dystrophin protein from being produced, while the milder Becker mutations still allow for the protein to retain some function. The symptoms for Duchenne, which affect 1 in 3,000 male births, start around 3-7 years old with difficulties in standing up and walking. This progressively worsens with most patients confined to a wheelchair by age 12 and dying of respiratory complications by age 20. Becker results in milder symptoms and a later age of onset with patients surviving into their 40s.

Alfredo (affectionately known as Dino) Ferrari was born in 1932 and suffered from Duchenne muscular dystrophy. Although leaving him very weak he continued to work with his father Enzo at his Ferrari car manufacturing company on the development of a 1.5 L DOHC V6 engine for F2, which was later renamed in his honour when he passed away at the age of 24. After his death, his father Enzo then founded the *Dino Ferrari Foundation* in

Milan, which is one of the most important research centres in the world for muscular dystrophies.

Enzo and Alfredo Ferrari.

Muscle cells, as well as nerve cells, require a great deal of energy, and are therefore particularly impaired by defects in mitochondria, the cell's power houses. Such mitochondrial myopathies, which can be caused by mutations in mitochondrial DNA, generally result in progressive muscle weakness that can disrupt the body's autonomic functions that are not consciously controlled, like breathing and heart disturbances. There are many different types of mitochondrial myopathies so far identified, with a complex array of symptoms. Some symptoms can be so mild that they are hardly noticeable, while others are life-threatening. Matthew "Mattie" Stepanek, the US child poet, suffered and died from a very severe mitochondrial myopathy shortly before his 14th birthday. He began writing poetry in early childhood in order to cope with the death of his older brother, who along with another brother and sister, also died of the same disease at a young age. His five volumes of verses sold millions even making the New York Times' Best Seller list. In addition, he served as the goodwill ambassador for the Muscular Dystrophy Association, speaking at events across the US. At his funeral in June 2004, a eulogy delivered by President Jimmy Carter at his funeral included the following sentence: "*We have known kings and queens, and we've known presidents and prime ministers, but the most extraordinary person whom I have ever known in my life is Mattie Stepanek.*" As Mitochondria are only inherited maternally, it was Mattie Stepanek's mother who carried and passed on the mutant mitochondria. She only suffered from a mild adult form, diagnosed after her children were born. A technique is currently being developed to allow mothers, carrying mitochondrial mutations, to conceive unaffected children. This involves transferring only the nucleus from a fertilised egg, which may contain defective mitochondria, from a very early human embryo into an unfertilized egg, containing fully functional mitochondria, from another woman.

Mattie Stepanek.

Not all muscular diseases involve muscular degeneration. Myotonic disorders (Gr. *myo*; muscle, *tonic*; strength), which affect around 1 in 15,000 individuals, are characterised by muscles which can contract normally but have decreasing power to relax. These diseases result from a block in the flow of electrical impulses across the muscle cell membrane; without the proper flow of charged particles the muscle cannot return to its relaxed state after it has contracted. Generally, this results in a painless muscle stiffness in the limbs and, typically, an inability to release the handgrip rapidly. These diseases usually have no effect on a person's lifespan. One form, known as Thomsen's myotonia congenita, is caused by inheriting a mutation in a gene affecting the function of a chloride channel that regulates the flow of ions across the membrane in cells and is a key component in a variety of biological processes involving rapid changes in cells such as in muscle contraction. It is named after the Danish physician Dr Julius Thomsen who, in 1876 provided the first documented account of a myotonic disorder which he discovered in himself! He was supposedly highly sensitive about his familial condition, and it was only in response to accusations that the youngest of his three affected sons was trying to avoid military service, that he was provoked into publishing his article describing how he could trace back the condition to his maternal great-grandmother born in 1742. There is a breed of domestic goat with a similar myotonia congenita. Known as the fainting goat, their muscles freeze for roughly 10 seconds when the goat is startled, causing the animal to collapse on its side.

Another class of Myotonic disorders is myotonic dystrophy (MD). Effecting around 1 in 8,000 this autosomal dominant disorder can lead to development of a mask-like expressionless face, typically with eye drooping, though symptoms can vary considerably. It has been suggested that the pharaoh Akhenaton, of the Eighteenth dynasty of Egypt, had MD. Reigning from around 1350 BC artistic representations of him portray a strikingly bizarre appearance, with a long face, thin and hollow cheeks, a half-open mouth and lowered eyelids known as ptosis. The most common type of MD is caused by a trinucleotide repeat expansion mutation in a gene called the Myotonic Dystrophy Protein Kinase (DMPK), which produces a regulatory protein important in a number of cellular pathways. The reason that eyelid drooping is a characteristic sign of myotonic dystrophy, in addition to a number of muscle weaknesses disorders, is due to the fact that the muscle that raises the upper eyelid is particularly sensitive to fatigue. This is why people in general experience heavy eyelids when tired. Some people, however, are born with a specific weakness affecting only this muscle in either or both of the eyes and so are unable to raise the eyelid completely. Called congenital ptosis, this can be genetically inherited or acquired, and while the incidence worldwide is still unknown, a number of celebrities appear to have this condition such as Salman Rushdie and the British singers Gabrielle, and Thom Yorke of *Radiohead* fame. What is interesting though, are the relatively large numbers of actors and actresses with ptosis, perhaps most famously, Forest Whitaker who won best actor Oscar for *"The last king of Scotland"*. Film critics often refer to Whitaker's droopy left eye, attributing to it the actor's enigmatic sleepy, contemplative look. Another Hollywood star, Sylvester Stallone, also owes much of his trademark sneer, slurred speech and ptosis to a partial facial paralysis that he was born with.

Bust of Pharaoh Akhenaten.

Forest Whitaker

In contrast to gene defects leading to muscle loss or weakness, some genes, when altered, can lead to increased muscle strength and growth. Belgian Blue cattle, also known as *monster cows*, contain a mutation in the myostatin gene resulting in a shortened myostatin protein, which is unable to function in its role of limiting muscle growth. As a result they contain a third more muscle than normal cows producing large amounts of very lean meat. One child was recently born in Germany inheriting the human form of the same mutated gene and was consequently born with excessive muscle growth in the arms and legs. The boy had inherited two mutated myostatin genes allowing his muscles to grow unregulated. Analysis had shown that a mutation, from each parent, ran in the family. This myostatin gene was first discovered by scientists in a particular strain of mutant mice, subsequently named *mighty mice*, that had twice as much muscle as normal. It appears that dog breeders have been inadvertently selecting for this same gene for many years in whippet racing. A study showed that a high number of dogs in the top racing classes had one copy of this mutant gene. Dogs that inherit two copies of the mutant gene are particularly muscular and are known in the whippet-breeding community as *bully whippets* which, contrary to the name and appearance, have a very placid temperament.

It is possible that the myostatin protein might be of great therapeutic value in humans if it can signal muscle growth in individuals suffering from muscle wasting diseases such as muscular dystrophy. Some laboratories have already developed drugs that can block myostatin function in mice. In addition to myostatin there seem to be many other gene mutations that can lead to increased muscle growth and strength. One gene known as ACTN3 codes for a protein found only in fast twitch muscle fibres. Professional sprinters tend to have a high percentage of fast twitch fibres in the skeletal muscle helping to produce the explosive bursts of speed and power, while marathon runners conversely tend to have more slow twitch fibres. Non-mutated highly functional version of this ACTN3 gene has been found to be significantly more common in professional sprinters and fast twitch fibre abundant individuals than in endurance athletes who tend to contain a mutated version of this gene. This seems to suggest that the determination of muscle fibre is to some degree hereditary.

A Bully Whippet.

Motor neurons originate in the spinal cord or brain stem and connect with muscle fibres to allow muscle contraction. For muscles to function, grow and survive they must be constantly stimulated by the motor neurons; if not, the muscles weaken and waste away. Diseases resulting in the death of motor neurons result in a progressive paralysis of the body, while leaving the mind intact. The legendary baseball player, Lou Gehrig, died from a motor neuron disease known as Amyotrophic Lateral Sclerosis (ALS) (Gr. *a;* absence, *myo;* muscle, *trophic;* nourishment). This is caused by degeneration of nerves in the lateral areas of the spinal cord which are replaced by hardened tissue, a process known as sclerosis. This leads to progressive weakness, often starting in the hands and feet and proceeding up the arm or leg, and also spasticity where muscles twitch, become tight and spasm. Often known as Lou Gehrig's disease in America, ALS affects about 1 in 30,000. Over 90% of cases seem to appear at random for which a primary cause is still not properly understood, although the disease has been linked to several factors. However, of those cases that do show an hereditary component some are caused by mutations in a gene known as SOD1, which normally serves to clear free radicals from cells. Different mutations in this gene affect the age symptoms begin and how fast the disease progresses. Some sufferers first experience symptoms later in life followed by rapid deterioration of function and death within a couple of years after diagnosis, while for others the first appearance of symptoms occurs at an earlier age but progression is much more gradual.

David Niven first began to notice symptoms of ALS at the age of 70, dying 3 years later from the disease in July 1983. He began to notice changes in his speech and physical ability; his arms and legs would sometimes begin to ache, and his voice would began to develop a faint slur. This slur became particularly evident during an interview for the BBC in 1981 with many viewers assuming that he was drunk. It was a few months later that he was diagnosed with ALS, though he continued to write his next novel, and filmed "*The Curse of the Pink Panther*" and "*The Trail of the Pink Panther*". However, he appeared cadaverous in these films, and his voice had to be dubbed. After a brief period of hospitalisation, Niven returned to his chalet at Chateau d'Oex in Switzerland, where his condition continued to decline until he passed away.

David Niven 1967.

In contrast to David Niven, Stephen Hawking developed the symptoms of ALS at the very young age of 21 and has been living with the disease for over 40 years. When he was 17 years old at Oxford University he was rowing, but by his third year at Oxford he noticed that he seemed to be getting clumsier, and fell over once or twice for no apparent reason. The disease progressively worsened until the age of 40, from which time he has needed 24 hour nursing care. Underlining the fact that motor neuron diseases have no effect on mental abilities, Stephen Hawking has held the post of Lucasian Professor of Mathematics at Cambridge University since 1979, continuing to combine family life (he has three children and one grandchild), and his research into theoretical physics together with an extensive programme of travel and public lectures.

While the risk factors for ALS are not properly understood, head and spine trauma has long been suspected. This is suspected from the unusually high proportion of ALS sufferers among Italian soccer players from the top three Italian divisions from the 1960s to 1996 (33 out of a total of 24,000). It has been suggested that this occurs from heading footballs, or the use of neurotoxic pesticides on soccer fields, in genetically predisposed individuals.

Stephen Hawking, 1999.

Affecting around 1 in 6,000 people, the most commonly occurring motor neuron disease is Spinal Muscular Atrophy (SMA). Even though it is five-times more common than ALS, it receives little publicity due to the fact that most affected people die very young, whilst awareness of ALS has been raised by high-profile celebrity sufferers. The SMA disorders are usually autosomal recessively inherited and are actually the most common cause of genetically determined neonatal death. Caused by different mutations affecting the Survival Motor Neuron (SMN) protein, as many as one in 40 Europeans are carriers. There are three main types of spinal muscular atrophy. The acute type usually results in death within the first year, while a less severe form presents between the ages of 5 and 15 years with a comparatively slower progression of weakness. However, as with most genetic diseases of this type, rate of progression of symptoms and life expectancy are variable with some sufferers surviving into adulthood. Ami Ankilewitz long outlived his predicted life expectancy. Working as a 3D animator in Israel, at the age of 36 he created and stared in an award winning documentary, "*39 Pounds of Love*", chronicling his pursuit of a lifelong ambition - an American road trip to find the doctor who predicted he would not live more than a few years.

Chapter 6

DERMATOLOGICAL DISORDERS

Tigers die and leave their skins; people die and leave their names.

Japanese Proverb

Supposedly, the largest person ever to have walked the earth was a farmer named Mills Darden who lived in North Carolina in the early 1800s. Estimates suggest he stood 228.6 cm tall and weighed up to 490 kg. So sensitive was he about his size, he never consented to be weighed and it was only through a clever scheme of his neighbours, who wished to know his weight, that an estimate was obtained. They marked the exact point the springs on his horse-drawn cart lowered to as he sat on it and then later, when Mills was not about, loaded large rocks on the cart to match it. From this we can determined the surface area that his skin would have covered using the formula of height [cm] x weight [kg]/ 3600)1/2. If this estimate is accurate it would give Mills a skin surface area of nearly 2½ championship sized snooker tables (15.55 m^2), if it were stretched out flat! In contrast, the skin of an averaged sized adult would cover about 1.82 m^2 meters.

The skin is made up of an outermost epidermis (Gr. *epi*; on, *derma*; skin) and the dermis. Below this is the subcutaneous layer that contains fat cells, serving to conserve heat and protect internal organs from physical damage. Skin thickness varies on different parts of the body. The palms and soles of the feet (1.2 mm to 4.7 mm) have the thickest skin while the thinnest skin is found on the eyelids and lips (~0.05 mm). Males generally have thicker skin than females and it is the Rhinoceros that has the thickest skin of any terrestrial mammal with some areas as thick as 2.5cm.

The epidermis is composed mainly of keratinocyte cells and melanocyte cells, while the dermis underneath contains blood vessels, nerves, hair follicles, muscle, glands and lymphatic tissue. Each square cm of human skin can consist of up to four million cells, 24 hairs, 35 oil glands, 6.1 meters of blood vessels, 246 sweat glands, 7,480 sensory cells, 23,622 pigment cells, and more than 393 nerve endings. The major cell type in the epidermis are keratinocytes (Gr. *cyte*; hollow - nowadays translates as a cell). These cells produce strong structural proteins called keratins (Gr. *Keras*; horn), that enable the cell to keep a robust rigid structure, and are also the main component of nails, hair, feathers and horns. There are over 30 different types of keratin proteins each produced by a different gene. As the keratinocytes develop, they move up the skin layers producing different types of these keratin proteins forming rigid scaffolds in the cells. As the cells reach the layers of the stratum corneum they die forming the dead cell layer of our outermost skin. This layer forms a barrier against the entry of foreign material and infectious agents, and minimizes moisture loss from the body.

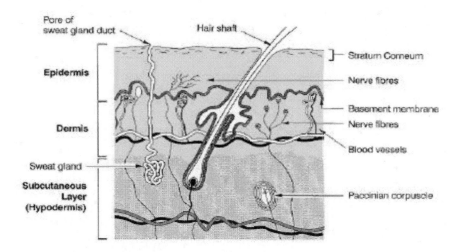

Diagram of a cross-section of the human skin.

The production of keratin in our skin is quite dynamic and those areas of skin on our body that begin to need more protection, perhaps due to constant rubbing or pressure, will stimulate the local production of more keratin. Known as hyperkeratosis, this results as the formation of protective calluses, such as found in the hands of farm labourers, or corns on the feet. Some people are born with an inherent massive overproduction of skin keratin leading to very thick and tough scaly skin. Caused by mutations in keratin genes these conditions are known as ichthyoses (Gr. *ichthys*; fish) as the hardened skin can sometimes vaguely resemble the scales of a fish. Ichthyosis vulgaris (Lat. *vulgaris*; common) is the most frequent form of ichthyosis and one of the most widely inherited skin disorders, affecting around 1 in 1,000 people. It is not life-threatening and symptoms often consist only of mild itching. Usually dominantly inherited, this condition is caused by the lack of a structural protein called filaggrin, which is produced in the epidermis. While inheriting two copies of a mutant filaggrin gene results in ichthyosis vulgaris, the presence of one copy of the mutant gene, which up to 10% of individuals in the UK have, can lead to a dry flaky skin in the form of eczema. One man famous for his ichthyosis was Emmitt Bejano, whose calloused skin allowed him to give public performances where he would sit submerged in vats of ice water. Known by his stage-name of the *Alligator-Skinned Man*, he made the headlines when he eloped with Priscilla Lauther, another side-show performer known as the *Monkey Girl*, who had a disorder, described later, which caused hair to grow all over her face. Priscilla's foster father disapproved of Emmitt Bejano, and so when the two ran off together to marry in 1932, a local newspaper famously penned the headlined: "*Monkey Girl Kidnapped by Alligator Man.*" They remained together until Emmitt Bejano died in 1995. They had only one child who died not long after birth. Many sufferers from this condition have often appeared as *Alligator* or *Lizard* people in various sideshows, owing to the sometimes unusual looking skin and, although it was the Englishman Robert Wilan who first made an accurate description of ichthyosis in 1808, references to the condition have been found in ancient medical texts dating back more than 2,000 years.

John H Williams, *The Alligator Boy,* circa 1930.

A more severe type of ichthyosis is lamellar ichthyosis (LI). Babies born with the disorder are often encased in a thick membrane which soon dries and peels off, leaving the baby with bright red underlying skin which over time develops into large, brown scales. Affecting 1 in 50,000-100,000 people, this recessive disorder is caused by mutations in a gene for an enzyme involved in forming the structure of the epidermis. Another form of the disorder results in heavily uneven skin with the appearance of ridges or spikes developing from the thickened skin. Known as ichthyosis hystrix (Gr. *hystrix*; porcupine) one of the first people to be described with this condition was *The Porcupine Man,* Edward Lambert. Born in Suffolk in 1717 to parents without the skin anomaly, he earned his living by exhibiting himself and his one affected son in travelling circuses. Subsequent generations of the family who inherited the condition carried on the family trade of exhibiting themselves around Europe.

Different mutations in genes coding for some types of keratin proteins can lead to extremely fragile skin. This occurs in epidermolysis bullosa (EB) (Gr. *lysis*; release/loosening, Lat. *bulla* – bubble/blister) where an afflicted person's skin is so fragile, that it can tear or blister from what is normally routine activity or contacts, such as wearing clothes or warm temperatures. This autosomal dominant disease, affecting around 1 in 50,000, results from keratin proteins that are less able to support the stability of keratinocyte cells. Dependent on which keratin proteins are affected, skin fragility can vary from the mild EB simplex to the severe dystrophic EB. This latter condition affected Newcastle born Jonny Kennedy who was the subject of the film documentary *"The Boy Whose Skin Fell Off"*, detailing the final months of his life. One scene shows his mother changing the protective bandages for his blisters that covered three-quarters of his body. As she does this his skin comes off with the bandages no doubt causing excruciating pain. A tireless campaigner for recognition of the disease, the documentary he inspired, broadcast in March 2004, attracted millions of viewers. *"I like to see it as the skin is like Velcro,"* he explained, *"the hooks are missing. So, any knocks or severe friction just doesn't go well with the condition, because it pulls the skin off.*" In 2003 he was diagnosed with skin cancer, for which there is a high risk factor associated with this disease, and sadly died in September of that year on a train returning from a meeting at 10 Downing Street with Cherie Blair. However, typing his name into a web search will reveal the phenomenal scale of awareness Jonny Kennedy brought to EB and the support he generated for the EB charity DebRA.

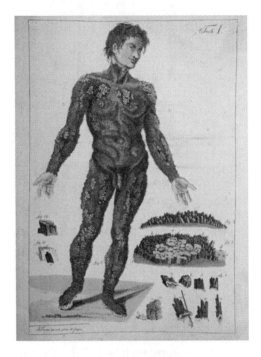

Edward Lambert, *The Porcupine Man*. Tilesius, 1802.

Jonny Kennedy.

Keratinocyte cells originate from the division of stem cells in the dermis where they move through the epidermis until flaking off from the skin's surface to make room for the new cells coming from deeper layers. This process takes about 28-30 days and the human skin generally sheds itself at a rate of about a million cells every 40 minutes. However, this process of skin shedding accelerates in conditions of excessive keratinocyte proliferation. One such disease is psoriasis, which affects around 2–3% of the population, and results from hyperactivity of the immune system leading to inflammation and a rapid turnover of skin cells. In monozygotic (identical) twin studies there is a 70% chance of a twin suffering from the condition if the other twin already suffers. A number of different genes, which regulate the immune system, are suspected of causing the disease leading to an autoimmune reaction. The British author and playwright Dennis Potter suffered from Psoriatic Arthropathy, a condition that, in addition to affecting the skin, also causes arthritis in the joints. This began in the early 1960s and led to frequent hospitalisation. He was sometimes unable to move without great pain. His fingers also became immobile so that he could only write by strapping a pen to his hand. He depicted his disease in one of his most famous plays "*The Singing Detective*".

Dennis Potter.

The skin has various associated structures such as hair, nails, teeth and sweat glands. The blueprint for the development of these are laid down during embryonic development, starting with the formation of thickenings of the ectoderm (the ectoderm is the outermost of the three layers (the other two being the mesoderm and endoderm) that make up the very early embryo) called placodes. The inability to properly form these placodes can lead to diseases referred to as Ectodermal Dysplasias (ED) (Gr. *ecto*; outside, *derma*; skin, *dysplasia*). These encompass many related genetic disorders that result variously in missing teeth, sparse hair, inability to perspire, malformed nails and cleft lip/palate. Around 1 in 1,500 people are affected by a form of ED. The most common form is X-linked Hypohidrotic ED (Gr. *hypo*; reduced, *hidrotic*; sweating), caused by mutations in a gene on the X chromosome coding for a protein controlling the development of ectodermal tissue. This condition affected the 'toothless men of Sind' who lived in Hyderabad in India and were described by Darwin in 1875: "*..ten men, in the course of four generations, were furnished, in both jaws taken together, with only four small and weak incisor teeth and with eight posterior molars. The men thus affected have* *very little hair.... suffer much during hot weather from excessive dryness of the skin. It is remarkable that no instance has occurred of a daughter being affected... though the daughters transmit the tendency to their sons; no case has occurred of a son transmitting it to his sons".* The American actor Michael Berryman was born with this disorder, resulting in him lacking hair, fingernails, teeth and sweat glands. The millions of sweat glands on the average human body produce moisture to help keep the body cool, so the lack of sweat glands in Michael Berryman forces him to rest often to avoid overheating. He was actually a florist before Hollywood director George Pal noticed him, and his physical characteristics have been used to great effects in a number of horror movies. Interestingly, there is also a breed of bald, toothless dog, known as the *Mexican Hairless Dog*, supposedly used by the Aztecs as a type of bed warmer, which contain defects in the same gene.

Michael Berryman, 2007.

As well as developing into the skin and its appendages, the ectoderm also gives rise to the nerves, brain, and spinal cord. Other defects in the development of the ectoderm in an embryo underlie a group of disorders known as neurocutaneous syndromes. These are characterised by skin abnormalities, usually appearing as skin lesions, and can lead to the development of tumours affecting the central nervous system. The most common of these syndromes are neurofibromatosis and tuberous sclerosis.

Neurofibromatoses (NF) are autosomal dominant disorders characterised by the development of a type of tumour, known as a neurofibroma, from the cells and tissues covering nerves. These benign (not malignant) skin tumours can cause bumps under the skin and coloured spots called café-au-lait spots as they resemble milky coffee. In addition they may also affect the bone and nervous system leading to skeletal problems and other neurological conditions. However, the severity is highly variable ranging from skin lesions in some individuals to severe disfigurement in others. The underlying cause for the development of these tumours is linked to increased stimulation of nerve growth due to mutations in genes coding for proteins that inhibit cell growth and division known as tumour suppressors. The two major types are NF-1 (often known as Von Recklinghausen's disease or peripheral NF) and NF-2, which is referred to as central NF. NF-1 is the most common type affecting 1 in

3,000-5,000, while the incidence of NF-2 is quite rare with 1 case per 50,000-120,000 people. The depiction of Quasimodo is strongly suggestive of NF-1. He was described as having cysts on his skin and lumps and bumps associated with neurofibromas. He also had a curved spine and a large head, which can be further symptoms. From the accurate descriptions of the disease in the book, it has been suggested that Victor Hugo, the author of the story, might have known someone with the disease. Quasimodo's disfigurement led to hostility from the ignorant townsfolk and forced him to lead a life of seclusion in the bell tower of Notre Dame in Medieval Paris. It is this same ignorance that has driven the Peruvian born Gabriela Bazan to push for further awareness of this disease. In 2004, at 24 years of age, she started losing the hearing in one of her ears and began experiencing frequent headaches. She was eventually diagnosed with NF-2. The main problem of NF2 is the development of multiple tumours on the brain and spinal cord, and while not having as many outwardly signs as NF1 one of the first symptoms of NF2 is the growth of tumours on the auditory nerves that lead to hearing loss. Shocked by the lack of awareness of neurofibromatosis in Peru Gabriela Bazan set about informing her fellow countrymen about the disease by appearing regularly on television, and travelling across Peru giving public talks, establishing foundations, and serving as an inspiration to many.

Another neurocutaneous syndrome resulting in benign tumours developing in the brain and the skin is tuberous sclerosis. The widely respected French physician Désiré-Magloire Bourneville first described this disease in 1880, using the name "*sclerose tubereuse*" to describe the occurrence of characteristic tuber or root-like growths in the brain arising early in brain development and becoming calcified and hardened (sclerotic), over time. The disorder affects around 1 in 6,000-10,000 individuals, and although it may be present at birth, signs of the disorder can be subtle with full symptoms taking time to develop. As a result this disorder is frequently unrecognised and misdiagnosed. As with NF1 and NF2, tuberous sclerosis is also caused by mutations in various genes which code for proteins that normally regulate cell growth, i.e. tumour suppressors.

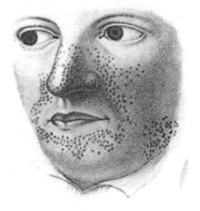

Tuberous sclerosis.

Often confused with NF1 is Proteus syndrome (*Proteus*; Greek God of the sea with an ability to change shape). This condition causes skin overgrowth, abnormal bone development, and tumours on the body. Proteus syndrome is an extremely rare disease, with only 100-200 individuals in the world affected, although it is now suspected that there are many more milder cases. It is a progressive condition whereby children are usually born without any obvious deformities but as they age tumours, in the form of skin and bone growths, appear. Some studies suggest a role for a particular tumour suppressor gene and familial cases of the disease do appear to exist. However, very little is still known about this disorder and it is uncertain whether or not the condition results from a mutation occurring during embryogenesis. In this case, only a portion of cells would contain the defective gene (i.e. mosaicism). Joseph Merrick, known as the *Elephant Man*, is suspected to have suffered from Proteus syndrome. There are some suggestions that he might have had a sister with a similar condition. He was born in 1862, to a teacher called Mary Jane in a poor area of Leicester. Growths started erupting on his skin by the age of two and his disfigurement worsened as he grew older with his head growing larger and his right arm becoming severely deformed. Abandoned and cheated out of all his money Merrick was discovered getting off a train Liverpool Street station by the well-known doctor, Frederick Treves. He was astounded that the young man was not only highly intelligent, but also literate and a great lover of poetry, and he petitioned Queen Victoria to arrange for a small apartment for him. Money was donated to buy new clothes and, suddenly, he became socially accepted having friends in the highest levels of London society. However, Merrick had always to sleep sitting up with a mass of pillows behind his back as his head was so heavy that if he lay down his windpipe would be crushed by the weight. But one night in 1890, he removed the pillows from his bed before he slept and was found dead the next morning – asphyxiated at the age of 28. Although upset by his death, Frederick Treves wasted no time in having Merrick's body boiled down to the skeleton which is still displayed in the Royal London Hospital museum along with his mask, oversized hat, and a model church he had made.

Joseph Merrick, 1890.

'Tis true my form is something odd,
but blaming me is blaming god.
Could I create myself anew,

I would not fail in pleasing you.
Was I so tall, could reach the pole,
or grasp the ocean with a span;
I would be measured by the soul.
The mind's the standard of the man.

- poem by Isaac Watts that Joseph Merrick would use to end his letters.

The skin is the largest sensory organ of the body containing millions of nerve endings known as sensory receptors. Motor neurons send information away from the central nervous system to stimulate cells such as muscle cells, whereas sensory (or afferent) neurons, send information from the sensory receptors, such as found in the skin, joints, muscles, organs, eyes, nose, tongue etc., towards the central nervous system. Sensory neurons can indicate pressure, temperature and pain and, when stimulated, transmit a signal to nerve cells in the spinal column which can either elicit a quick response independent of the brain, or can convey the message to further parts of the brain which can then send back an appropriate response to specific motor neurons to move various parts of the body. In some inherited disorders, such as congenital insensitivity to pain with anhidrosis (CIPA), these sensory neurons are disrupted rendering individuals insensitive to sensations such as pain and temperature. Sufferers injure themselves in ways that would normally be prevented by feeling pain and temperature, such as cuts and burns. However, it is overheating, due to an inability to control body temperature and sweating, which kill most children with the disease before the age of 3. One documented case is of a young Canadian student who attended University in Montreal. Physicians concluded that Ms. C was normal in every way except that she felt no pain. Pinching her tendons, injecting histamines under the skin, and, bizarrely, inserting sticks up her nostrils, all failed to produce the pain the doctors were looking for. Even when they subjected parts of her body to electric shock, hot water at burning temperatures or a long ice bath, no changes in blood pressure, heart rate, or respiration occurred. Her condition led to severe problems with her knees, hips and spine because of the stress to which they were subjected by abnormal flexure. Ordinarily the sensation of pain protects joints from this. Another side effect of the ability to regulate body temperature is that patients can be more at risk to infections and are unable to control fever. Sadly, Ms. C died at the age of 29 of massive infections, though during her last month she reportedly complained of pain and discomfort, which her physicians were unable to explain.

I have a dream that my four little children will one day live in a nation where they will not be judged by the colour of their skin but by the content of their character.

Martin Luther King, Jr.

Skin colour is chiefly the result of differing amounts of the pigment melanin (Gr. *melas*; black). This forms in little packets called melanosomes in a special cell called the melanocyte. These cells are found in the skin, eye and hair follicles and contain many branches which are used to pass the melanosome pigment packets they produce on to keratinocyte cells. These keratinocytes then store the melanosomes over their nucleus where the pigment serves to absorb harmful ultraviolet radiation preventing it reaching the DNA of the cell and causing mutations.

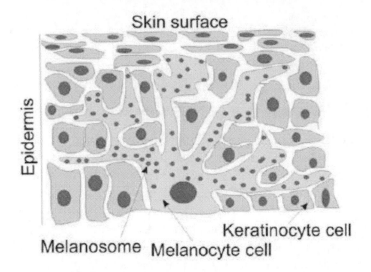

Cross-section of skin showing a melanocyte cell containing packets of melanin pigment called melanosomes.

Although everybody has roughly the same numbers of melanocyte cells, there is a great variability in the number and size of the melanosomes produced by these cells. Darker-skinned individuals produce larger and more numerous melanosomes and consequently rarely develop skin cancer and so are better adapted to tropical climates. However, in colder environments, with less sunlight, lighter skin is more advantageous as it allows the person to absorb more sunlight for vitamin D production. There is also evidence that individuals with heavily pigmented skin are more susceptible to frostbite. This was first noticed during the Korean War. A study showed dark-skinned soldiers were four times more likely to suffer from frostbite than those with lighter-skin.

Pigment disorders can be caused by mutations in a variety of genes leading to defects in either: 1, the production of melanin; 2, formation of melanosomes; or 3, the development of melanocytes.

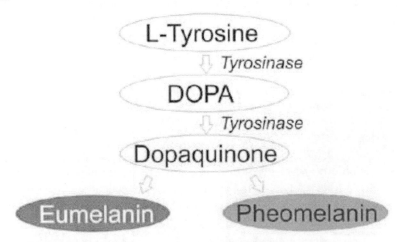

Basic pathway for the production of eumelanin and pheomelanin.

Defects in the production of the melanin pigment lead to a group of disorders known as albinism (Lat. *albus*; white) resulting in little or no pigment in the eyes, skin, and hair. Approximately one in 17,000 people have one of the various types of albinism, most of which are recessively inherited. The term albino was first used by the Portuguese in the 1500s to describe, what they originally thought was a new race of light-skinned Africans. In the book of Enoch, Noah is described at the time of his birth as having a body white like snow and hair white as wool suggesting that he could be the first documented case of albinism.

Andy Warhol, 1977.

There are two types of melanin; the dark coloured eumelanin (Gr. *eu*; true) and the red-yellow pheomelanin (Gr. *Phaios*; brown/dusky). Both of these are formed from the amino acid tyrosine in a biochemical pathway involving a number of steps where one compound in a pathway is converted to the next compound by the action of an enzyme. Mutations in the gene for the tyrosinase enzyme can result in an absence of both types of melanin. Depending on the mutation and the degree to which tyrosinase activity is disrupted, various types of oculocutaneous (Lat. *oculus*; eye, *cutis*; skin) albinism (OCA) occur. A completely inactive tyrosinase enzyme causes an albinism classified as OCA1A where individuals are born with white skin and hair, and blue eyes, which remain throughout life. They are also unable to tan (i.e. produce melanin) in response to sun stimulation. Photographs suggest that the musical brothers, Edgar and Johnny Winter, may have both inherited this form of albinism. Mutations that only reduce but do not completely remove tyrosinase activity result in a milder form known as OCA1B. People having this condition show very little pigment at birth but develop yellow hair (due to a slight production of pheomelanin) in childhood, and slowly accumulate pigment in their eyes, and skin as they age. Some mutations can leave the tyrosinase enzyme temperature-sensitive, so that it does not work above 35°C. This leads to the production of melanin only in cooler areas of the body, such as arms and legs, while warmer parts of the body, such as under the arms and the scalp, remain white. Some photographs of Andy Warhol seem to suggest he may have had this. Other enzymes, working downstream of tyrosinase in the melanin synthesis pathway, can affect the production of only one of either eumelanin or pheomelanin. For example individuals with OCA3 are only able to produce the red pheomelanin and therefore have a reddish skin, ginger/red hair, and brown eyes. This is quite common in the South African population where it is known as "Rufous" or "Red OCA". In the first book of the Bible a description of Esau suggests that he may have been born with this form of albinism. *And the first came out red, all over like a hairy garment; and they called his name Esau,* Genesis 25:25.

In contrast to mutations affecting melanin production, other forms of albinism occur as a result of mutations in genes involved in the functioning of the melanosomes inside the melanocyte cells. This occurs in OCA2 where individuals lack a protein involved in transporting molecules, such as the tyrosinase enzyme, into and out of melanosomes. This is the most common type of OCA and is particularly prevalent in equatorial Africa. Caucasians with OCA2 show white to yellow hair, blue-grey eyes, and white skin that does not tan on sun exposure while Africans generally also show yellow hair and light brown skin colour. Other genes, producing proteins responsible for the production of melanosomes, when mutated, can lead to a type of albinism known as Hermansky-Pudlak syndrome. These genes, in addition to disrupting melanosomes, also lead to defects in the production of similar organelles known as lysosomes, which are used, among other things, to store blood clotting chemicals in platelets. Consequently, this form of albinism also occurs with mild bleeding and though extremely rare is more common in Puerto Rico where it occurs at a frequency of 1 in 1,800 and seems to be the result of a founder mutation. Bleeding disorders generally never appear with albinism except in Hermansky-Pudlak syndrome, which is why it is interesting that in the movie "*Cold Mountain*", there is a villain with albinism who, in addition to the white makeup and bleached hair, suffers frequent nosebleeds.

Salif Keita, 2006.

Considering the particular disadvantage of albinism to people in sunny climates, it is surprising that the occurrence of this condition is so high among Africans. In some populations in Zimbabwe, 1 in 1,000 people have albinism and around one in 35 Africans are carriers of an albinism mutation. However, people with albinism often have a particularly hard time in Africa. Albino women in Zimbabwe, for example, are increasingly the victims of rape, because of the mistaken belief among Zimbabwean men that sex with an albino woman is a cure for HIV. Cinema is also guilty of discrimination against individuals with albinism, constantly casting people with albinism as villains – directors use the condition as a devise to distinguish the villain from the hero by means of appearance. James Bond, for example, has dispatched a number of albino villains through the years. Unfortunately, the depiction of albinos as villains has had an overwhelmingly negative impact on how people with albinism are perceived in society.

Portrait of George Alexander Gratton, unsigned, 1811.

For a very long time, it was thought that another pigment condition known as Piebaldism (*pie*; as in the black and white magpie bird, *bald*; as in the bald eagle which has a white feathered head) resulted from one parent being an albino and the other being dark skinned. Characterised by distinct patches of skin and hair lacking pigment, piebaldism can often be identified by a white forelock of scalp hair. It is now known that this results from defects in the migration of special cells from the ectoderm, during embryonic development, to the skin where they divide to produce melanocyte cells. Mutations in some genes (usually the KIT protoncogene) can interfere with this migration resulting in some patches of skin lacking melanocytes. One such case is of George Alexander Gratton, who was known as the *Spotted Boy*. Born in 1808 on the Caribbean island of St Vincent, he was taken as a baby to work in an English circus where he was purchased by the well-known showman, John Richardson. He treated him like a son and was distraught when the child died, possibly at around 8 years old. He was buried, with much ceremony, at the All Saints Church in Marlow in the same vault in which John Richardson himself was later interred.

A further pigmentation anomaly occurs in incontinentia pigmenti where there is a loss of melanin from cells which then collects in the dermis, causing a characteristic marbled pattern of discoloured skin. However, other areas of the body such as the brain and eyes can also be affected. This disease is often caused by mutations in a gene called NEMO, an abbreviation of NF-kappaB essential modulator that is found on the X chromosome. Inherited as an X-linked dominant disorder, incontinentia pigmenti occurs mainly in females as males cannot survive without a functioning NEMO gene, and thus die *in utero*.

WHAT IS IN A NAME?

In contrast to many diseases, genes are usually not named after their discoverers, but are given names that try to convey something about the functions or significance of the gene. However, a name like NEMO might not have been wholly coincidental, and some researchers have displayed a quirky sense of humour when "choosing" a name for a newly discovered gene. Sonic hedgehog is perhaps one of the more famous, describing the human equivalent of a fruit fly gene. The fruit fly gene was named hedgehog because, when mutated, it resulted in fruit fly embryos being covered with hair like a hedgehog. So when the American scientist, Bob Riddle, discovered the human equivalent of this developmentally important gene, he decided to call it sonic hedgehog after the computer game. But there are boundaries to naming a new gene, as one research group in New York discovered in 2005 when calling a new gene they had found *pokemon*, claiming it was an acronym for POK (a previously named family of genes) erythroid myeloid ontogenic. However, it was soon noticed that this gene had a role in the development of human cancer and consequently attracted such tongue-in-cheek headlines as *"Pokemon's cancer role revealed"*. This led Pokémon to exert its legal right to the trademark threatening to sue the researchers if they did not stop calling the gene *pokemon*. Sadly, the researchers had to cave in and this gene is now referred to, much more forgettably, as Zbtb7. However, scientists have so far got away with naming the gene for the arylsulfatase E, ARSE, and the viral gene fuculokinase, FUCK.

Some disorders lead to a loss of pigment later in life, for example when there is a destruction of melanocyte cells. This occurs in vitiligo, the most common depigmentary disorder of the skin with around 1% of population suffering pale and light-sensitive skin. Though appearing at any age, the condition tends to peak in the second and third decades of life. Genetics does play some role in vitiligo with various studies indicating that up to 30% of patients who report the condition also occurring in other family members. However, it does not follow a simple inheritance pattern suggesting the involvement of a combination of multiple genes in addition to environmental factors such as emotional stress or severe illness. This disorder was first recorded in 1500 B.C. in the ancient Indian Veda manuscripts and, along with psoriasis, is thought to be what is referred to as leprosy in the bible rather than the current usage of the term to denote Hansen's disease. Though vitiligo is often unnoticeable on Caucasian skin, on darker-skinned people it is much more obvious. This condition affected Michael Jackson, and his father and paternal grandfather are both thought to also have had a milder form of the disease. As his melanin-free patches of skin were sensitive to the sun he would usually wear long sleeves, a hat and gloves when outside. Supposedly, the resulting patches of discoloured skin covered around 80% of his body prompting him to use makeup to even out his complexion.

Vitiligo appears to have formed the basis of a number of interesting stories through the ages. One such story is of a Black slave in the US named Henry Moore who lived in the 1700s and, in a similar way to Michael Jackson, turned completely white in a short space of time even developing blond hair. It is recorded that as a result of his skin change he was treated much more kindly the white population. They collected money to buy his freedom and he spent the rest of his life living as a white man. However, the more common cases of vitiligo in dark skinned individuals result in distinct white patches. In fact the word vitiligo is thought to derive from the Greek *vitelius* for calf describing the resemblance of the white spots of vitiligo to the white patches on a cow.

Michael Jackson.

Abnormally increased pigmentation (hyperpigmentation), in contrast to the aforementioned cases of hypopigmentation, occurs when there is either an increase in melanin, or in the number of melanocyte cells. Changes and boosts in pigmentation occur in many of us, particularly due to contact with solar radiation which stimulates an increase in melanin production, through the hormone melanotrophin. This is known as tanning. However, other than this, patches of hyperpigmentation exist on the skin of most individuals, either to local increases in melanin, the melanosome packets containing melanin, or the melanocyte cells. For example, while café au lait macules (which, although occurring in association with neurofibromatosis, are generally non-syndromic) occur from increased melanin, freckles are the result of an increased number of melanosomes, while lentigines occur due to increased numbers of melanocyte cells. However, the sudden development of patches of hyperpigmentation or irregularly pigmented lesions could be a cancer of melanocytes, known as a melanoma. These can be caused by extreme sun exposure, especially during childhood. The World Health Organisation estimate that as many as 60,000 people die in the world each year as a result of too much sun, usually from skin cancers.

Chapter 7

HAIR DISORDERS

Long on hair, short on brains.

French Proverb

Hair is formed in hair follicles, located in the dermis. New hair is made in the hair bulb, at the base of the hair follicle, where cells expressing high amounts of keratin proteins multiply and push upward. These cells rapidly die and compact into the dense, hard mass forming the hair shaft.

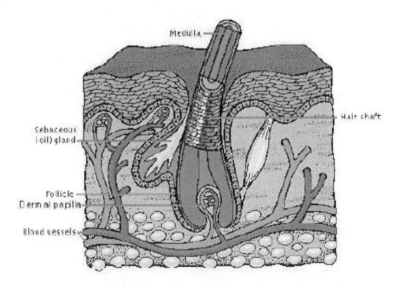

Diagram of a Cross-section of a hair follicle.

The human body produces three different types of hair shafts: lanugo, vellus and terminal hairs. Lanugo hairs (Lat. *lana*; wool) are found only on the foetus and are usually shed during the 7-8th month of foetal life. Vellus hairs (Lat. *vellus*; fleece) are short, soft hairs barely visible to the human eye, which cover most areas of the human body. This type of hair can often become more prominent in patients with anorexia where the lack of calories stimulates production of more of this type of hair in an attempt to retain body warmth. This also explains why many feral children, are often reported as being covered with body hair when found, which eventually disappears after they are returned to a better diet. Terminal hairs are the coarser often pigmented hairs generally found on the scalp, pubic area, armpits etc.

On average, there are 100,000 to 150,000 terminal hairs on the human scalp. Some hair loss is normal, and on a typical day, about 50 to 150 scalp hairs fall out of their follicles and are replaced by new hair. When these lost hairs are not replaced properly, balding occurs. Around a quarter of men begin balding by age 30 and two-thirds by age 60. Samuel Johnson explained it as *"the drying up of the brain and its shrinking from the skull"*. However, the vast majority of hair loss cases are the result of androgenetic alopecia (Gr. *alopex*; fox – an animal that frequently suffers a contagious scabies-like dermatitis, caused by mites, which leads to hair loss). As suggested by its medical name, androgenetic alopecia (also known as male pattern baldness) involves male hormones, known as androgens, and genetic factors. This is because the growth of terminal hairs is regulated to a varying degree by androgen hormones. At the base of the hair follicle is the dermal papilla which contains receptors for the androgens testosterone and, its more potent derivative, dihydrotestosterone (DHT). Androgens increase the size of hair follicles in areas such as the beard and underarm during puberty, but can conversely cause hair follicles in the scalp to decrease in size later in life. This was first noted from the fact that eunuchs, who no longer produce androgens due to castration, never seem to go bald. Androgenic alopecia obviously occurs far more frequently in men than in women, as women have low levels of androgens. However, women who acquire abnormally high levels of testosterone, for whatever reason, not only go bald, but also tend to grow beards. This is known as Hirsutism and may be congenital or acquired due to various diseases resulting in elevated androgens or else medications such as androgenic oral contraceptives. Although the exact causes of androgenetic alopecia are still not fully understood, it is hereditary to some degree and there are many genes which influence male pattern baldness. One in particular is the androgen receptor gene found on the X chromosome. One variant of this gene is often found among men who suffer balding at a very early age and is thought to result in more androgen receptors on the scalp making the scalp more sensitive to the effect of androgens and consequently leading to hair loss. Efforts to develop a cure for male pattern baldness are focussing on possible drugs that could inhibit DHT.

A more severe form of scalp balding is known as alopecia areata totalis where all scalp hair is lost in a relatively short time and at a relatively young age. Pierluigi Collina, the Italian football referee, suffered from this losing all his hair in the space of just two weeks when he was 24. Alopecia can also affect hair on other areas of the body, and if all body hair is lost, in addition to the scalp, the diagnosis then becomes alopecia areata universalis. This condition is shared by the British actor and comedian Matt Lucas, who lost every hair on his body around the age of 6, New Zealand Model Anna Fitzpatrick, and British swimmer Duncan Goodhew. Another condition, called diffuse alopecia areata, may cause a person with mixed white and dark hairs to lose only their dark hairs, giving the appearance of hair turning from grey to white very quickly, sometimes over less than a day. Supposedly this happened to both Thomas More in 1535 and Marie Antoinette in 1793 the day before each was executed. Emotional stress would seem to be a contributing factor in the occurrence of this phenomenon. However, most forms of hair loss are generally thought to be a type of autoimmune disease, in genetically predisposed individuals, where the body treats its hair as foreign tissue.

Pierluigi Collina.

Su Kong Tai Djin.

In contrast to baldness, some disorders exist that are characterised by too much hair, such as hypertrichosis (Gr. *hyper*; excess, *trichos*; hair). Hypertrichosis lanuginose is a genetic disorder resulting in fine light-coloured lanugo hair remaining on all parts of the body, except the palms and soles, which contain no hair follicles. Lanugo hair is the name given to hair grown on every person in utero and lost during pregnancy. However, in Hypertrichosis lanuginose, this hair is not lost and instead keeps on growing. This is usually dominantly inherited, though a gene for this has yet to be identified. A number of individuals with this disorder have been documented though the ages. One of the first recorded cases was of a boy called Petrus Gonzalez who worked in the court of the French King, Henry II. Born on Tenerife in 1556, he was covered from head to toe in blond hair. Given to the King as a gift, he proved to be highly intelligent and was quickly promoted at the court. At the age of 17 he went on to marry a French woman and raised several children, with the eldest son and two daughters sharing his condition. After King Henry died, Petrus and some of the family moved to a small village by Lake Bolsena in Italy, where he lived until the ripe old age of 80. Sadly, his two affected daughters suffered similar fates to a number of other people with the same disorder - being displayed like wild animals at circuses and sideshows. Known as *Lionel, the Lion-Faced Boy*, Stephan Bibrowsky was born in Poland in 1890 with lanugo hair covering his entire body. Moving to the USA he was one of the top circus attractions. It was claimed that his mother saw his father ripped to shreds by lions during her pregnancy resulting in his appearance. Fedor Jeftichew, who was born in St. Petersburg in 1868, toured with French circuses with his father who had this same condition. Given the stage-name of *Jo-Jo the Dog-Faced Boy*, the story was that a hunter found them in a cave as savages, and they could not be tamed. His act would consist of growling at the audience who were oblivious to the fact that Fedor was perfectly fluent in several languages. Not all individuals with this condition have ended up displaying themselves in various sideshows. Born in China in 1849, Tai Djin was abandoned in a forest to die by his superstitious family who took his condition to be a curse. However, a Shaolin monk travelling through the forest discovered the tiny child and took him back to the Shaolin Temple where Tai Djin was raised by the monks who taught him martial arts. Tai Djin proved to be exceptionally gifted and quickly become a favourite of many of the Shaolin masters who each passed their knowledge on to him. Consequently, Tai Djin became the first Grandmaster of Shaolin and the first to master all skills of the seven Shaolin temples, including over 200 different empty hand systems, over 140 weapon systems, and specialties such as the infamous *Chi Ma*, or *Death Touch*. Still revered around the world, he lived until the age of 79 and taught many others to be masters.

Julia Pastrana, 1900.

In addition to lanugo hair, pigmented terminal hair can also grow uncontrollably on the body in a condition known as hypertrichosis terminalis. One well documented case is Julia Pastrana, a Mexican woman who was born in 1854, with thick black hair covering her entire body. She married a man called Theodor Lent (in reality she was actually sold by her mother to him) who exhibited her around the world in various circuses as *The Ape Woman*. Paradoxically she was a highly intelligent woman who spoke several languages and was an accomplished musician. At the age of 26 she gave birth to a boy with the same disorder, though the child lived only thirty-five hours and Julia Pastrana herself died five days later. Lent promptly had his wife and his son's bodies mummified and carried on exhibiting them. Subsequently Theodor Lent met and married another woman with a similar condition to Julia Pastrana and he toured with her (under the pseudonym *Zenora Pastrana* – the name of Julia's sister) and the mummies. When Theodor Lent suffered a mental breakdown and died in an asylum, Zenora Pastrana continued touring with the mummies before selling these to the owner of a shop of oddities in Oslo. The mummies now reside in the Oslo Forensic Institute.

Hairy ears syndrome.

On certain parts of the body hair does not usually grow at all; when it does this abnormality may be linked to genetic mutations., An example of this is growth of hair on the outside of the ear. Though a relatively trivial condition, the inheritance of this anomaly has sparked much debate among geneticists through the years. In 1907 a father-to-son inheritance of this condition was first recorded suggesting that the gene responsible locates on the Y chromosome. This was later challenged as the Y chromosome, the smallest chromosome, was for some time considered to contain no genes. It is now known that the Y chromosome contains a number of genes that have attracted a lot of attention owing to the role of the Y chromosome in male sex determination. More recent studies have however found that the hairy ear syndrome may not be linked to the Y chromosome after all anyway and appears to be inherited in an autosomal dominant sex-limited pattern.

The shape and diameter of terminal hair varies widely among individuals. People of Asian descent have the thickest hair and on Caucasian scalps blond and black hair are thinner than red. The cross section of a human hair is usually round in people of Asian descent, causing the hair to be straight, while in Europeans this cross-section is round to oval, and in people of African descent, this cross-section is flattened, leading to the hair adopting a frizzy form. If a round cross-sectioned hair is flattened, such as by using the edge of a coin, it will also curl up into an afro-like style. There are a number of genes controlling this cross-sectional shape, and mutations in these genes can lead to the hair shaft exhibiting a flattened cross section. This can be seen in a number of disorders characterised by the occurrence of woolly hair, such as hereditary woolly hair syndrome, and the familial woolly hair which in addition can also be accompanied by heart abnormalities.

Woolly hair syndrome.

While hair is normally highly flexible, certain disorders are characterised by brittle hair that breaks very easily. Some of these weaknesses can result from defects in keratin protein formation. Sulphur is important for the structure of keratin proteins; it is this sulphur that gives the pungent odour when hair is burnt. Low levels of sulphur, such as occurring in trichothiodystrophy disease, results in weakened hairs. Keratin also contains the amino acid arginine, and low levels of this underlie Sabinas Brittle Hair syndrome. This autosomal recessively inherited disease, first observed in a small town in northern Mexico called Sabinas, results from a deficiency of an enzyme important for the production of arginine. A number of other enzymes involved in keratin formation rely on the presence of copper, and one disease resulting from low copper levels is Menkes kinky hair syndrome. This condition is often referred to as bamboo hair due to twisting of the hairs and the occurrence of tiny fragile nodes on the hair giving a bamboo appearance.

Hair colour, like skin colour, derives from the two types of melanin: eumelanin in black or brown hair and pheomelanin in auburn or blond hair. Albinism, as previously mentioned, presents with white hair due to defective melanin production. However, a progressive loss of melanin, due to decreasing numbers of melanocytes, is frequently seen in middle-aged people; a condition known as Canities (Lat. *canus*; grey), or Poliosis (Gr. *poliosis*; grey). Therefore, the process of greying hair is actually an illusion caused by the presence of white hairs, lacking melanin, among a percentage of normal dark hairs. The appearance of white hairs between the ages of twenty and thirty is known as premature greyness, and this can be inherited. One gene has been found to be essential for the maintenance of the melanocyte cells and mice lacking this gene rapidly turn grey shortly after birth. Therefore, it is suspected that some premature greyness sufferers may have a defective version of this same gene.

Elsa Lanchester as the *Bride of Frankenstein*, 1935.

Although white hairs can appear randomly on the scalp, mixing in with pigmented hairs, it is possible for white hairs to become concentrated at specific locations such as a white forelock. Dickie Davis had a memorable white forelock. Although this generally does not signify any underlying syndrome, a number of genetic disorders, such as Piebaldism and Waardenberg disease do tend to present with such a white forelock. This white patch of hair is often colloquially known as a witch's streak as, similarly to albinism, it also often associated with the caricature of witches and evil women in films.

Melanotropins are hormones that bind to receptors, such as the melanocyte-stimulating hormone receptor 1 (MC1R) on melanocytes, causing the cell to produce eumelanin. However, a lack of these melanotropins, particularly the previously mentioned alpha-MSH, result in the melanocytes not being stimulated and, instead, producing phaeomelanin leading to red/blond hair. This can also associate with obesity because as well as binding to the MC1R receptor, alpha-MSH also binds to the MC4R receptor in the brain that controls appetite. However, only a small proportion of redheads are overweight because the majority of individuals producing phaeomelanin do not contain mutations in the alpha-MSH gene, but contain alterations in the MC1R receptor gene. This MC1R receptor gene exists in many different mutant forms differing in their activities, which in a variety of combinations cause the different colours of hair from red, to auburn, to strawberry blond. A majority of the red-headed Celts have receptors that are almost totally inactive.

It is interesting that red hair is most commonly found in people of Scandinavian descent and it is thought that the clusters of redheads in the British Isles derive from Viking or Pictish ancestry. Scotland has the highest proportion of redheads of any country with 13% of the population having red hair and a further 40% carrying the highly inactive MC1R variant gene. It may be that the high occurrence or red-hair, particularly in northern Europe, is due to evolution selecting for red-headed people, because their accompanying lighter skin allows them to absorb more sunlight for vitamin D production.

Chapter 8

RESPIRATORY AND HEART DISORDERS

LADY BRACKNELL: . . . Do you smoke?
JACK: Well, yes, I must admit I smoke.
LADY BRACKNELL: I am glad to hear it. A man should always have an occupation of some
kind. There are far too many idle men in London as it is.

Oscar Wilde, The Importance of Being Earnest

The respiratory system, of which the lungs are the largest part, play a vital role in delivering oxygen to the body, removing carbon dioxide waste, regulating temperature and stabilizing blood pH balance. Within the lung system oxygen and carbon dioxide are passively exchanged between the gaseous environment and the blood, facilitating both oxygenation of the blood and the removal of carbon dioxide and other gaseous wastes from the circulation. To achieve this, the environment of the lung needs to be very moist. This, however also makes it a hospitable environment for bacteria and indeed many respiratory illnesses are the result of bacterial or viral infection of the lungs. To combat this, and to prevent tissues from drying out, we produce mucus which traps the foreign material preventing it from entering the lungs during breathing. This mucus mainly consists of mucin proteins and inorganic salts suspended in water, but also contain antiseptic enzymes and proteins involved in the immune system. In addition to its role in the lungs, mucus is also found in the digestive system, where it is used as a lubricant aiding the passage of food down the oesophagus.

Increased mucus production in the respiratory tract is stimulated in response to many diseases, such as the common cold. However, one disease called cystic fibrosis results in the mucus becoming too viscous, therefore impeding breathing. Cystic fibrosis (CF) - the name was originally used to describe a disease of the pancreas - is the most common autosomal recessive disorder in Europeans affecting around 1 in 2,500 births in the UK. In the Middle Ages there was a superstition that infants with salty skin were considered "bewitched" as they would routinely die an early death. It was a common practice to lick the forehead of a child to taste for the presence of salt. It was only many hundreds of years later that this observation was linked to how cystic fibrosis is caused. The disease results from a salt imbalance, whereby a mutation in the CFTR gene hinders the passage of chloride (which, along with sodium, makes up salt) into and out of the cells leading to dehydration and thickening of the extracellular mucus covering the cells lining the respiratory and digestive systems. This thick, sticky mucus clogs the lungs causing breathing problems and frequent lung infections, such as pneumonia. Furthermore, thickened digestive fluids made by the pancreas are prevented from reaching the small intestine where they are needed to digest food leading to affected people often showing slow weight gain and growth.

Frédéric Chopin, 1849.

Frédéric Chopin is thought to have suffered from CF. Born in Poland in 1810 he is widely regarded as one the greatest of composers for the piano. However, his frail health started early in childhood with recurrent diarrhoeas and gastro-intestinal ailments resulting in weight loss, and frequent respiratory tract infections. Chopin was often so weakened by his disease that he was unable to climb stairs and occasionally had to be carried out after performances. These symptoms progressively worsened, often confining him to his bed for long periods later in life, from where he would give his piano lessons in a lying position. Although records from the autopsy state that Chopin's death was caused by "*a disease not previously encountered*", it is very likely that this was CF. An hereditary basis to his disease seems to be supported by the fact that Frédéric's father had recurrent pulmonary ailments. Two of Chopin's sisters also suffered from respiratory problems with one dying at a young age. More recently the British singer-songwriter and pianist, Alice Martineau suffered and died from CF. Thinking that her condition would prevent her from singing, Alice did not at first pursue her musical ambitions but she later discovered that her constant coughing had actually strengthened her vocal chords. Releasing her debut album in 2002, she died shortly after at the age of 30. Before she died she took part in a documentary about CF that was broadcast in 2005. As the recessive CF affects around 1 in 2,500, it is possible to work out what proportion of people carry the mutant CFTR gene by dividing 1/2500 by a 1/4 chance of the two genes being inherited together, 1/625. The carrier rate of the CF gene then is 1/(square root of 625) giving a figure of 1 in 25 of the UK population carrying a single copy of the mutant CFTR gene. It is also possible to work backwards to calculate the disease incidence if the gene frequency is known; if a carrier rate of the mutant CFTR is 1/25, the number of couples at risk are the carrier rate in males x carrier rate in females; 1/25 x 1/25 = 1/625 couples at risk. The actual disease incidence can calculated by the taking into account the 1 in 4 chance two heterozygote carriers produce an affected child, i.e. 1/625 x 1/4 = prevalence of 1/2500.

This is an unexpectedly high proportion of mutant CFTR carriers considering that two copies of the gene confer such low life expectancy; it would be expected that evolution would have selected against the gene. The reason for its prevalence could be that having one copy of the gene provides some protection against cholera. This is known as an heterozygous advantage and occurs in a number of other diseases mentioned in this book.

Cholera produces a toxin that binds to the cells of the small intestine opening chloride channels and leading to large losses of chloride ions and water in the form of diarrhoea. If these are not quickly replaced an infected person will die of dehydration. Intestines of mice carrying one mutant CFTR gene when infected with cholera secrete considerably less water than mice with two functional CFTR genes. Therefore, it may be that humans carrying one copy of the CFTR gene when infected would extrude enough water to flush the intestines of the toxin without succumbing to the diarrhoea and dehydration. There is some evidence that the mutant CFTF gene also confers resistance to typhoid fever. The *Salmonella typhi* bacteria causing typhoid fever can only invade the gastrointestinal cells by attaching to the normal CFTR protein and not the mutant version.

In addition to mucus, as a further barrier against infections, the lungs also contain white blood cells which release a powerful enzyme called elastase which breaks down proteins. This enzyme has to be carefully controlled so that it only attacks invading pathogens and not the normal host tissues in the lungs. This would lead to emphysema where the small air sacs in the lungs become severely damaged. To combat this our lungs produce a second protein, known as alpha-1 antitrypsin (AAT), which is able to inactivate the elastase in the immediate vicinity of the delicate lung tissue. Between 1 in 1,500 and 1 in 7,000 people suffer from alpha-1-antitrypsin deficiency suggsting that the mutant AAT gene is as common as the mutant CFTR. This disease can lead to a shortness of breath appearing between the ages of 20 and 40 years and the development of emphysema (Gr. *emhysan*; inflate) caused by destruction of lung tissue.

Some toxins, such as cigarette smoke, directly inactivates AAT so speeding up the process of elastase induced lung damage. People with different mutations in the AAT gene can therefore have differing degrees of susceptibility to toxins such as tobacco. This genetic link might have been responsible for the pattern of smoking related illnesses in the family of Richard Reynolds. He was one of founders of the cigarette industry in the US and his family company still manufactures four of the United States' 10 best-selling cigarette brands. After years of smoking his own product, Reynolds died of emphysema, followed by two of his four children, and many subsequent family members who also smoked heavily. This phenomenon of inherited susceptibilities to certain toxins has been a contentious issue in a number of court cases. One of the first was in the 1970s and involved the Dow Chemical Company who had started testing prospective employees for AAT deficiency. The rationale was that industrial pollutants, in the same way as cigarette smoke, might hasten the onset of lung disease in AAT deficient individuals. The company claimed that it was trying to protect prospective employees found to be AAT deficient. The union argued that the testing was an unfair labour practice and maintained that the onus should be on the industry to make the workplace safe for everyone, regardless of their AAT status. The company subsequently discontinued the tests.

Cilia are tiny hair-like structures found lining the windpipe where they constantly beat in one direction sweeping mucus and dirt out of the lungs. These cilia are also found in the oviducts, where they help move the ovum from the ovary to the uterus, and also in the tails of sperm. Defects in cilia can cause a number of diseases such as primary ciliary dyskinesia and polycystic kidney disease. Primary ciliary dyskinesia, also known as immotile ciliary syndrome, is a rare autosomal recessive genetic disorder caused by a defect in cilia, leading to a loss of cilia movement. The result is reduced mucus clearance, and susceptibility to respiratory infections such as bronchitis and pneumonia. Interestingly, many patients also experience hearing loss and a poor sense of smell due to the role cilia play in olfactory function. Infertility is also common in males with this condition owing to the role that cilia play in sperm mobility.

In 1643 an old soldier who died in battle under the command of Louis XIV, ended up on the autopsy table under what was thought to be a routine confirmation of death. He had not fathered any children and reportedly had a poor sense of smell. When cutting up the body the surgeon Marco Severino was amazed to discover that the old man had his heart on the right side of his body; other organs were also positioned on opposite sides to normal. What Marco Severino had discovered (a peculiarity first reported a century earlier by the Scottish doctor Matthew Baillie) was a condition, common among people with inherited cilia defects, known as situs inversus (Lat. *situs*; location, *inversus*; inverted). This is due to another important role that cilia play, in foetal development. They cause a current of molecules, which signal the development of particular organs, to flow across the embryo resulting in different concentrations of these on the left and right side. Impaired cilia function can result in a loss of this current and consequently half of all patients show situs inversus with, most commonly, the heart on the right-hand site and the liver on the left. About one in 8,500 has this, usually autosomal recessive abnormally from mutations in a number of different genes affecting cilia. Contrary to what it may seem, this does not generally result in any medical problems for the person concerned and indeed, some have become professional athletes, like Randy Foye, a basketball player for the Minnesota Timberwolves.

Once I had brains, and a heart also; so having tried them both, I should much rather have a heart.

The Tin Woodsman of Oz

The heart is a muscle that pumps blood around the body. It is divided into four chambers: right/left atria and right/left ventricles. The atria are thin-walled chambers that receive blood from the veins, while the two ventricles are thick-walled chambers that forcefully pump blood out of the heart. Four valves keep the blood flowing in only one direction: travelling from the heart to the lungs where it picks up oxygen, after which it is then pumped through the body by way of arteries. As the oxygen is used up by the body's tissues and organs it is returned by way of veins to the heart, where it is once again pumped to the lungs.

Congenital heart defects, describing those present at birth, are the most common birth defects - affecting almost 1 percent of all babies born - and are the leading cause of birth defect-related deaths. Congenital heart defects include a variety of malformations of the heart or its major blood vessels that are present at birth. These defects may obstruct blood flow in the heart or vessels, or cause blood to flow through the heart in an abnormal pattern. Ventricular Septal Defect, affecting the wall dividing the left and right ventricles, is generally considered to be the most common type of malformation, accounting for around a third of all congenital heart defects. While the cause of most congenital heart defects is unknown, some heart malformations do have a clear genetic link. Genetic causes of heart disease include chromosomal numerical abnormalities such as trisomies 21, 13, and 18, large chromosome aberrations such as 22q11.2 deletion syndrome, and single gene defects as seen in syndromes such as Noonan syndrome. In addition, a vast and diverse number of genetic syndromes also present with heart defects as one of the symptoms.

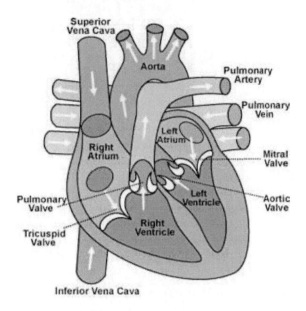

A diagram of the heart.

Noonan syndrome (NS), affecting around 1 in 1,000-2,500 individuals, is one of the most common conditions associated with congenital heart anomalies. This autosomal dominant condition, named after Dr Jacqueline Noonan, a paediatric cardiologist still working in Kentucky, results from mutations in gene encoding a signalling molecule important in regulating cell growth and differentiation during development. The main characteristics, in addition to heart defects, are short stature, learning problems, indentation of the chest, impaired blood clotting, and characteristic facial features. It has been suggested that the blacksmith in the famous painting "*Among Those Left*" by Ivan Le Lorraine Albright had Noonan syndrome, judging from the contour of the breastbone, the low-set ears, and the short height. Most families with this disorder carry a mutation in a gene important in embryonic development, particularly in the formation of the heart valves, which serve to prevent blood flowing back from the arteries into the ventricles.

Among Those Left. Ivan Le Lorraine Albright, 1929.

In contrast to single gene defects, 22q11.2 deletion syndrome is a disorder caused by the deletion of around 3 million base pairs of DNA near the middle of chromosome 22. This region contains about 30 genes which when missing result in symptoms including heart defects and an opening in the roof of the mouth known as a cleft palate. With an incidence of around 1 in 1,800 births, the features of this syndrome are so varied that it has been given a number of different names such as DiGeorge syndrome and Shprintzen syndrome by researchers who were convinced that they had discovered a new syndrome. However, once it became clear that these symptoms were all determined by the same chromosome 22 deletion, they were grouped into a single syndrome. It then became known for a while as CATCH22 (for cardiac abnormality, T-cell deficit, clefting and hypocalcaemia, chromosome 22). Although a clever title, it has negative overtones and so fell out of favour. It now known more prosaically as the 22q11.2 deletion syndrome, which can be tested for, in a similar way to Williams syndrome, using Fluorescence *in situ* hybridization (FISH) (see *Cell Division and Chromosome Defects*). One heart abnormality often associated with 22q11.2 deletion syndrome is Tetralogy of Fallot, although a number of single gene mutations also seem to cause this. Affecting between 3 and 6 children in every 10,000 births and accounting for 10 percent of all congenital heart diseases, Tetralogy of Fallot is a complex of anatomic abnormalities arising from the maldevelopment of the right side of the heart. The well-known snowboarder Shaun White was born with Tetralogy of Fallot, for which he underwent two open-heart operations when he was one year old. The Dutch painter Dick Ket was believed to have had this with numerous self portraits showing another characteristic, often seen in this disease, which is a progressive enlargement of the fingers, known as clubbing, due to reduced levels of oxygen reaching these parts of the body.

Dick Ket. Self-portrait, 1939.

In addition to the major structural heart defects diagnosed at birth, as many as a further 1 in 150 babies in the UK are born with heart defects which often go undiagnosed initially because they are not associated with any other accompanying symptoms. Hypoplastic left heart syndrome, for example, though rare, is one the most serious types of congenital heart disease usually being fatal within the first few days or months unless it is operated on. Linked to mutations in at least three different genes it results in the left side of the heart being underdeveloped leading to restricted blood flow. Sudden infant death syndrome (SIDS), also known as cot death, occurs in around 1 in 8,400 children, usually under the age of 6 months. Although, the causes of most cases of SIDS are not clear, underlying heart defects appear from recent research to have a role in a significant percentage of cases. The genetic link explains why cot deaths occur at higher frequencies in certain families. Over 200 women in the UK in the last 20 years have been convicted of the murder or mistreatment of 1 or 2 of their infants who died without known cause, i.e. SIDS. The British paediatrician, Sir Roy Meadows, who was involved in the prosecution of many of these women, famously asserted *"one cot death is a tragedy, two is suspicious and three is murder"*. He calculated there to be a 1 in 73 million probability that a mother could have three children dying of cot deaths and this was a major factor in persuading juries to arrive at a guilty verdict in the trials of these women. This calculation completely ignores any possible genetic links which would greatly reduce the probabilities of cot death in any one family. It also ignores environmental factors such as bacterial infections, mould spores, and having a parent who smokes, which would also greatly reduce probabilities in any one family. Many of these unsafe convictions have since been overturned.

One particular abnormality, that studies suggest affects up to a third of infants dying of cot death, is a prolongation of the QT interval in the first week of life. The QT interval is the length of time it takes the electrical system to recharge following a heartbeat. This defect is also found in adults and accounts for a number of sudden adult death syndrome cases, described as the adult version of cot death in infants.

On June 26th 2003 in the 72nd minute of the Confederations Cup semi-final between Cameroon and Columbia, Cameroon midfielder Marc-Vivien Fóe collapsed and died in the centre circle. No other player was in five yards of him. The autopsy suggested he had long QT syndrome making an individual more vulnerable to a very fast, abnormal heart rhythm in which no blood is pumped out of the heart. In some cases there are underlying genetic mechanisms and a number of different familial long QT syndromes have been identified, with one mutant gene for a potassium channel (potassium ions are necessary for function of the electrical conduction system of the heart) underlying around half of all genetic cases.

Although there is little information about the number of young athletes who die from adult death syndrome, an Italian study suggested an occurrence of in 1 in 50,000 athletes a year. Over a period of 10 days in the summer of 2007 three footballers died suddenly, sending shock waves around the sporting world. On August 16th Leicester City player, Clive Clarke, suffered a heart attack during a League Cup tie at Nottingham Forest, though he recovered in hospital. Four days later Walsall's 16-year-old youth-team player Anton Reid suddenly collapsed and died on the training ground. Then on the 28th August the Seville left-back, Antonio Puerta, died after suffering a heart attack followed leass than 24 hours later by Chaswe Nsofwa, a Zambia international, who died of heart failure during a training session with his Israeli club side. Later in the year the Motherwell captain and midfielder Phil O'Donnell collapsed and died on the way to hospital of a seizure during his team's 5-3 win against Dundee United.

While long QT syndrome has been identified in a number of cases, the leading cause of sudden death in people under the age of 30, is an inherited heart condition known as hypertrophic cardiomyopathy (HCM). This results in the accumulation of an abnormal protein inside heart muscle cells that cause the muscles to thicken and the heart to develop an irregular beat. Presently, there are 9 genes identified that when mutated can cause the disorder; these code for filament proteins found in muscle cells of the heart. Probably about 1 in 1,000 to 2,000 people are born with HCM through all racial groups of which it is estimated that about 1 percent die from the condition. Most cases of HCM go undetected. Daniel, the son of former Wales's football manager Terry Yorath, died from HCM during a kick-about with his father in their back garden. He was 15 at the time and was embarking on a promising football career having just signed for Leeds United. Miklos Feher, Benfica's 24-year-old striker, was already well into his footballing career before HCM claimed his life during a 2004 league clash against Vitoria Guimaraes.

Excessive exercise can be a trigger for sudden death in otherwise healthy people with undetected heart defects and this has prompted a number of sport regulatory bodies to install compulsory medical checks for athletes. World football's governing body, FIFA, made sure every player at the 2006 World Cup was checked for possible heart defects. The Italians have taken a particularly strong stand on this issue and have passed laws requiring all athletes participating in active sports to undergo annual medical examinations before being allowed to take part in competitive events. Anyone found to be at risk of heart problems is automatically excluded. The Italian swimmer Domenico Fioravanti, who won two gold medals at the Sydney Olympics in 2000 for the men's 100-metre and 200-metre breaststroke, was forced to retire at the age of 27 after being diagnosed with HCM.

Chapter 9

BLOOD DISORDERS

Blood is inherited and virtue is acquired.

Venezuelan Proverb

The average person contains about 5 litres of blood, which is composed of about 60% liquid plasma (mainly water) and 40% cells. All the different types of blood cells are manufactured from stem cells (see *Neurological Disorders*) found in the bone marrow. These stem cells divide and differentiate into red blood cells called erythrocytes (Gr. *erythros*; red), white blood cells known as leukocytes (Gr. *leuko*; white), or platelets which are also known as thrombocytes (Gr. *thrombo*; clotting).

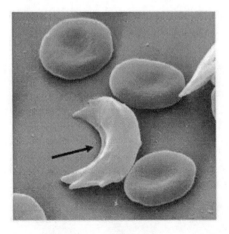

A sickle-shaped red blood cell.

Mammalian red blood cells (RBCs) have a flattened shape with a depressed centre (i.e. biconcave) optimizing the cell for the exchange of oxygen with its surroundings. These cells are able to carry oxygen due the presence of haemoglobin which is a complex molecule composed of globin proteins, and haem groups whose iron molecules temporarily link to oxygen. RBCs are also flexible so as to fit through the tiny capillaries in the body. In fact the average RBC travels around 1,000 miles through the maze of the body's circulatory system during their 120 days lifespan before being broken down in the spleen.

Defects in the shape of RBCs can lead to anaemia (Gr. *a*; absence, *haem;* blood) describing a deficiency in the function, or a decrease in the number, of RBCs. There are a number of gene mutations which can that result in this. The most common is sickle cell anaemia where RBCs adopt a sickle-like shape and lose their elasticity so increasing their chances of getting stuck in capillaries leading to downstream tissues becoming deprived of oxygen. This can, as a result, lead to periodic painful attacks, eventually leading to damage of internal organs or a stroke.

Sickle cell anaemia is an autosomal recessive disorder caused by a specific mutation in a gene encoding one of the β-globin chains making up haemoglobin protein complex. The disorder was first clinically described in 1910 by Chicago doctors James B. Herrick and Ernest Edward Irons who found "peculiar elongated and sickle shaped" cells in the blood of Walter Clement Noel, a 20 year old dental student suffering from anaemia. Although he was readmitted to hospital several times during his degree, Noel completed his studies returning to Grenada to practice dentistry. However, African medical literature actually first reported this condition in the 1870's when it was known locally as *ogbanjes* (*children who come and go*) because of the very high infant mortality. In fact, sickle-cell anaemia is the most common genetic disorder among Africans with about 1 in every 12 being a carrier. In addition to African countries the disease has a high occurrence in many other countries where there are high incidences of malaria. This is due to heterozygote carriers (i.e. individuals with only one mutant gene and so not suffering from sickle cell anaemia) showing some resistance to malaria. These individuals, with only one copy of the mutant gene, still produce a small number of sickle RBCs - not enough to cause anaemia, but enough to give resistance to malaria. The malaria parasite, transmitted by mosquitoes, spends part of its life cycle in RBCs which, in heterozygous sickle-cell anaemia carriers, results in the RBC rupturing leaving the parasite unable to reproduce. Therefore, heterozygotes in this area of the world have a greater evolutionary advantage; this is another example of a heterozygote advantage.

Tionne Tenese "T-boz" Watkins.

The incidence of sickle cell anaemia has been dropping among African Americans, due to the low levels of malaria in the US, but is still relatively common. One high profile sufferer is

Tionne Tenese "T-Boz" Watkins, the lead singer of the group *TLC*. She has been in and out of hospital intensive care units, since she was 7 years old with one of her longest hospital stays being for four months in 2002. "*Sickle cell is very excruciating pain,*" she says. "*It feels like somebody is stabbing you over and over again*". **Named one of** *People* magazine's 50 Most Beautiful People in 2000, she is one of the spokespeople for the Sickle Cell Disease Association of America. In the UK and the US all babies are screened for sickle cell anaemia in addition to phenylketonuria and a number of other disorders of metabolism, within the first 2 weeks of birth as part of the standard "heel-prick test" used to collect drops of blood for DNA.

Mutations in another gene can lead to RBCs adopting a small spherical shape, known as Hereditary Spherocytosis (*sphero*; spherical). This results in the cell contracting to its smallest and least flexible configuration - a sphere, rather than the more flexible biconcave disk. These cells have less surface area to which oxygen can attach, but the main problem is that these misshapen, but healthy, RBCs are mistaken by the spleen for old or damaged cells and are thus constantly being broken down by the spleen. This leads to anaemia and chronic fatigue. The standard treatment for spherocytosis is to remove the spleen.

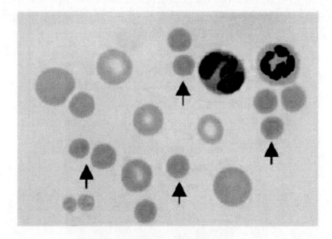

Spherocytotic red blood cells.

The globin part of the haemoglobin molecule is encoded by a number of different genes for alpha and beta globin. Sickle cell anaemia results from one specific mutation in the beta-globin gene, leading to a change in haemoglobin and cell shape. However, other diseases known as thalassaemias occur due to mutations leading to deficiencies in either the alpha- or beta-globin proteins making up the globin component of haemoglobin. This collection of autosomal recessive conditions can result in varying degrees of anaemia and are classified according to which chain of the globin molecule is affected; in alpha-thalassemia, the production of alpha-globin is deficient, while in beta-thalassemia the production of beta-globin is defective.

Unlike most genes, there are actually two genes for alpha-globin on each chromosome: the more of these genes that are mutated, the more severe the disease. Defects in all four genes (two on each chromosome) results in early death. Three affected genes leads to the development of anaemia in childhood. Two affected genes can cause a mild anaemia, and if only one of the four alpha genes are affected then no symptoms are usually seen. There is only one gene for beta-globin and if an individual inherits mutations in both parental copies then a severe anaemia called beta-thalassemia major results which, if left untreated, results in death before the age of twenty. This is sometimes known as Cooley's Anaemia, after American physician Thomas Cooley who first described it in 1925 and who, unusually among scientists, actually resented the disease being named for him. If an individual inherits only one mutant beta-globin gene from a parent, beta-thalassemia minor results. This is a mild anaemia which may show some symptoms of fatigue, but in most cases is asymptomatic with many people unaware they have this disorder.

The name thalassemia translates from Greek as "*anaemia by-the-sea*" (Gr. *thalassa*; sea) because at one time it was believed that the disease affected only people of Italian or Greek descent. This is now known not to be true as it is also common in Africa, the Middle East and Southeast Asia. The reason it is found in these regions is because, as in the case of sickle cell anaemia, carriers of the disease have a degree of protection against malaria. Pete Sampras, the son of Greek immigrants, has thalassemia minor, the symptoms of which sometimes troubled him during long hot-weather matches. After the Corretja match in the 1996 U.S. Open, a newspaper reported that Pete had Thalassemia. However, he did not admit it until he broke the Grand Slam record in Wimbledon 2000 because he didn't want his opponents to know that he was playing with a handicap! Zinedine Zidane, the son of Algerian parents also suffers from this, which causes nothing more serious than profuse sweating during games.

Pete Sampras.

RBCs are dependent on blood glucose as a source of energy through a biochemical pathway called glycolysis. The enzyme Glucose-6-phosphate dehydrogenase (G6PD), which is present in all human cells, is an important enzyme in this pathway and deficiency of this enzyme is the most common inherited enzyme deficiency disease in the world. Being found on the X chromosome, the vast majority of people with the condition are male. The condition results in an anaemic response to certain foods and chemicals such fava beans (broad beans) and the antimalarial drug, primaquine. This is because G6PD also plays a role in protecting red blood cells against oxidative damage – fava beans contain extremely high amounts of oxidants which can damage G6PD-deficient cells.

G6PD deficiency is probably the first example of a pharmacogenetic disorder to be described. Pythagoras, in about 500 BC in southern Italy, recognized that some individuals who ate the fava bean became ill, suffering from symptoms of anaemia including jaundice and fatigue, while others could enjoy them with no adverse effects. This leads to the commonly used term "favism" for this disease and Pythagoras used to call on his disciples to abstain from beans. This has however also been interpreted as Pythagoras exhorting his followers to stay away from politics; in ancient Greece and Rome, beans were used as a method of casting votes. Alternatively he may have observed that the high sugar content of beans leads to flatulence! By the late 1940s it was realised that the British, in contrast to Mediterranean populations, rarely developed haemolytic anaemia (an increased destruction of blood cells) on ingestion of fava beans, suggesting a genetic cause. This became even more evident during the Korean War when around 10% of male American soldiers of African or Mediterranean descent developed haemolytic anaemia in response to the antimalarial drug, primaquine, which was later discovered to be also due to G6PD deficiency. This was dramatized in an episode of "*M*A*S*H**" in which *Cpl. Maxwell Klinger* exhibited all the symptoms of malaria but did not have the parasite. However, when he stopped taking primaquine, he recovered completely.

The reason G6PD deficiency is particularly prevalent in people of Mediterranean descent – affecting as many as 1 in 10 – appears again to be due to the protective effect this condition has against malaria. The malaria-causing parasite, plasmodium, requires G6PD for its survival and replication in RBCs. It may be that G6PD-deficient RBCs infected with the Plasmodium, are cleared more rapidly by the spleen, or else that high numbers of oxidants in the blood of G6PD deficient individuals are lethal to Plasmodium. Since the mortality rate for favism is low and the mortality rate for malaria is high, the protective effect of the enzyme deficiency is significant. There are some villages in Sardinia where up to 70% of males are found to be G6PD deficient, possibly reflecting the high infant mortality rates from malaria in these areas after World War II, following deliberate flooding of some coastal areas by departing German troops. G6PD deficiency has also been connected to long life spans (longevity); in centenarians the incidence of G6PD deficiency is, on average, double compared to control groups. Some epidemiological studies have even suggested that G6PD may decrease susceptibility to cancer, cardiovascular disease and stroke.

Porphyria's love: she guessed not how
Her darling one wish would be heard.
And thus we sit together now,
And all night long we have not stirred,
And yet God has not said a word!

Porphyria's Lover, Robert Browning.

Haemoglobin, as previously described, is composed of globin proteins combined with a molecule called haem. This consists of an organic ring structure called a *porphyrin* and an iron atom which is able to bind oxygen. In addition to its role in haemoglobin, haem also has other functions serving as a component of several liver enzymes.

There are 8 steps involved in the making of haem from the 2 molecules glycine and succinyl-CoA, and each step is performed by a specific enzyme encoded by a different gene. If one of the different haem precursor molecules (known as porphyrins) is not modified, it cannot proceed to the next step, instead accumulating in various tissues. The diseases caused by partial defects in these enzymes (a complete absence of haem production is not compatible with life) are known as porphyrias (Gr. *porphyros*; purple), deriving from purple/red urine colour when the porphyrins are excreted. Affecting around 1 in 20,000 individuals, these porphyrins can accumulate in other parts of the body resulting in different physical and neurological symptoms, dependent upon which enzyme is affected and which porphyrin accumulates as a result.

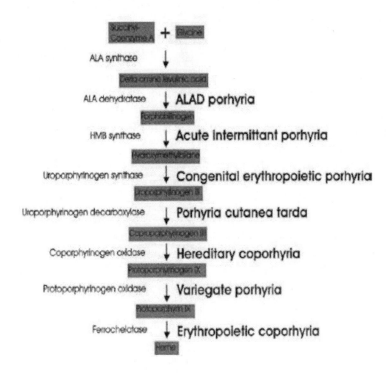

The haem biosynthetic pathway.

The most common inherited porphyria is acute intermittent porphyria (AIP), an autosomal dominantly inherited reduction of PBG deaminase, the third enzyme in the chain, resulting in the accumulation of the porphyrins delta amino levulinic acid (ALA) and porphobilinogen. Both of these are neurotoxins that can cause neurological complications. Though most people who inherit this trait never develop the symptoms, i.e. the disease shows a low penetrance, it is more common in psychiatric patients. Symptoms include abdominal pain, tremors, seizures and also psychoses, including agitation and hallucinations, which can last for weeks and then quickly abate. Precipitating factors for such attacks include hormones, drugs and alcohol which induce the production of haem in the liver. It has been suggested that Vincent van Gogh might have suffered from attacks of AIP, though the medical history of van Gogh could be a book in itself as there are many theories to explain his various ailments, and his artwork! There were, however, several bouts of severe psychosis in his adult life provoking suicide attempts and self-mutilation and necessitating institutionalisation. Factors triggering his symptoms could have included overwork, fasting, malnutrition, and of course his famous absinthe binges, explaining why, during hospitalization, these symptoms generally abated with improved diet and lack of alcohol. However, there was persistent depression throughout his life and in July 1890, at the age of 37, van Gogh walked into nearby fields and shot himself. Wounded, he returned to the Ravoux Inn, in Auvers, where he died two days later; his dying words were: *"La tristesse durera toujours"* – the sadness will last forever.

Defects in the forth step of the pathway lead to congenital erythropoietic porphyria (CEP), also known as Gunther's disease. In this autosomal recessive disorder, the porphyin, hydroxymethylbilane, builds up causing severe skin blistering and photosensitivity as the porphyrin absorbs light causing a burning sensation. Teeth can be reddish/brown or even fluorescent due to the porphyrin collecting in the enamel. Many have suggested that cases of CEP in the past could have formed the basis of vampire legends. Symptoms can include: photosensitivity of eyes and skin leading to patients avoiding daylight, severe blistering leading to scaring of the face, teeth enamel stained a red colour, stretching of the lips causing teeth to become more prominent, hypertrichosis leading to a lanugo-like hair growth over the face, and muscle weakness especially in the wrists and fingers. Interestingly, garlic stimulates haem production and so can turn a mild case of porphyria into a very painful one. It has even been suggested that people suffering from porphyria in past centuries might have been prescribed blood to drink in the mistaken belief that this could reduce the anaemia. Furthermore, although porphyrias are rare, it is believed to have been endemic among the Eastern European aristocracy, which was characterised by a high incidence of intermarriage. It is also recorded that Bram Stoker, before writing *"Dracula"*, met a psychiatric patient who exhibited many of the symptoms (including photosensitivity) associated with CEP, and this could have helped to inspire his novel.

Bela Lugosi as *Dracula*, 1931.

Variegata porphyria (VP) is an autosomal dominant porphyria caused by mutations in the gene for another enzyme in the haem production pathway. Although rare, it is common in South Africa where most cases can be traced back to a Dutch man, named Berrit Janisz, who carried the mutant gene, and emigrated there in the late 1600s. The term variegate describes the variable nature of the symptoms; most people do not experience any symptoms while others show a similar condition to that seen in AIP, with abdominal pains, skin rashes, and psychosis in addition to some photosensitivity. King George III, often called the mad king, is suspected of suffering from VP. Between 1788 and 1804, he suffered bouts of "madness" involving a rapid pulse, fever, abdominal pains, constipation, cramp, skin blistering, "*port-wine-coloured*" urine, and rambling speech degenerating into obscenities and hallucinations. Just as each episode was acute in onset, the recovery phase was equally sudden and rapid. However, in 1810 he descended into permanent derangement; deemed too mad and confused to rule, he was confined in Windsor Castle until he died in 1820. It has been suggested that these episodes of psychosis dramatically altered history. His disorder would have impaired his ability to make crucial decisions regarding, for example, the administration of the American colonies. Indeed, he was once seen to shake hands with a branch of an oak tree with the mistaken impression that it was the King of Prussia! George III's porphyria can be traced back to his ancestors, Mary Queen of Scots in the 1500s, and her son, King James I, whose physician noted urine "*red like Alicante wine*".

Portrait of King George III. Allan Ramsay, 1762.

Porphyria, like Haemophilia, is another hereditary disease that plagued European royal families. Other notable royal sufferers are said to include James I's son Henry, Prince of Wales who had a similar illness and died suddenly, with allegations that he had been poisoned, the Duchess of Orleans who died of a sudden episode of excruciating abdominal pain, Queen Anne, George IV, and Princess Charlotte who died suddenly after childbirth. Prince William of Gloucester (the current duke's elder brother, who was killed in a plane crash in 1972) had been diagnosed with porphyria suggesting the presence of a gene defect in the family and though it is possible that Queen Victoria may have also carried the gene, no symptoms were reported. Erythropoietic protoporphyria (Gr. *protos*; first), which is due to a reduced levels of an enzyme involved in the last step of the pathway converting protoporphyrin IX into haem, leads to very severe photosensitivity, with affected infants screaming soon after being taken out into the sun. It has been suggested that a builder involved in the construction of the Paris Opera house, who lived in the cellars to possibly to avoid sunlight, may have suffered from this disorder inspiring the French writer Gaston Leroux to create *"The Phantom of the Opera"*.

Molly Nash and her brother Adam.

RBCs are produced from stem cells in the red bone marrow of large bones - a process called erythropoiesis. Diseases resulting in a disruption of this bone marrow can lead to anaemia. One example is Fanconi anaemia, named after the Swiss paediatrician Guido Fanconi, which is due to an altered bone marrow growth predisposing individuals to myeloid leukemia. Failure of the bone marrow to produce blood cells – RBCs, white blood cells and platelets - in this disorder results in a susceptibility to infection, fatigue and haemorrhaging. There are several different types of Fanconi anaemia caused by mutations in different genes involved in the repair of DNA damage and therefore the protection of the body against cancer. With around 1 in 300 people carrying at least one of the mutations the occurrence of this disease is around 1 in 360,000 although the carrier rate for this mutation is much higher in Ashkenazi Jews at 1 in 90.

Parents of a girl called Molly Nash, born with Fanconi anaemia, underwent a controversial procedure to conceive another child who could help cure their daughter. They firstly underwent *in vitro* (Lat. *vitro*; glass, referring to experiments conducted outside of the body) fertilization, known as IVF, a procedure where egg cells are fertilised outside the woman's body and grown to the 6 or 8 cell stage and then transferred back to the mother. The embryos were genetically tested for signs of the FA causing gene and only an embryo which did not carry the genetic abnormality, and therefore a safe donor match for Molly, was then implanted into her mother. She subsequently gave birth to a healthy boy named Adam and blood from his umbilical cord containing stem cells was infused into his sister. Both children are fine, and Molly is now able to lead a normal healthy life. This procedure has raised fears that it could one day be used to screen for trivial genetic traits unrelated to medical defects, such as eye or hair colour, to produce so called designer babies.

PRENATAL DIAGNOSTICS

There are a number of techniques which are used to diagnose prenatal defects in foetuses whose mothers are at risk of having a baby with an abnormality.

Amniocentesis involves taking a sample of amniotic fluid (the 'waters') using a needle at 15-18 weeks. This can be analysed to determine certain biochemical, chromosomal or neural tube defects such as spina bifida, and can be used to test for specific genetic defects known to be inherited in the family. Another procedure is placental tissue (chorionic villus) sampling, which can be performed earlier at 10 to 12 weeks, but is considered to be more risky to the embryo than amniocentesis.

The most widely used technique is ultrasound scanning using ultrasonic waves to scan the foetus allowing visualisation and enabling skeletal and neural tube defects, among other abnormalities, to be identified. Scans are normally performed between 16 to 20 weeks, but since no actual tissues are sampled the technique cannot include genetic testing.

The practice of prenatal diagnosis has recently come under scrutiny, mainly due to different interpretations of what a "serious handicap" is. Many babies in the UK have been, and are being, aborted at advanced pregnancy stages with only mild defects that could easily be corrected, such as polydactyly, club feet and cleft palates. In some countries, notably China, abortion of healthy female babies, though illegal, is not uncommon, leading to a growing gender imbalance in the Chinese population; figures from 2009 show that for every 100 girls born, 119 boys are born.

Kristi Yamaguchi

The term 'club foot' often sounds worse than it really is, and is actually one of the most common birth defects in Britain, affecting around 1 in 500-1,000 newborns. The condition describes the presence of a foot pointing downwards and inwards, often as a result of conditions in the uterus. Although this can lead to a limp, the condition can usually be corrected in early childhood without any lasting problems. Indeed the condition seems to have had little effect on the footballing career of the American Mia Hamm who won Gold in women's soccer in 1996 and has the accolade of having scored more international goals than any other player male or female. Arguably one of the most talented footballers who ever lived, the Brazilian, Garrincha, was born with very disfigured legs; a right leg bent inwards and a left leg also bowed and 6cm shorter. In addition the American football stars Dan Marino, Troy Aikman, Charles Woodson and LeRoy Butler were all born with club feet. LeRoy's club feet were so bad that doctors had to break his legs and reset them while he was a toddler, forcing him wear bulky braces on both legs; 20 years later he would be named in the NFL 1990's All-Decade team. There are certainly many similar stories, such as that of the Japanese figure skater, Kristi Yamaguchi, who won Olympic gold in 1992 after only starting the sport in childhood as therapy for her club feet, or the exploits of Tom Dempsey who still holds the record for the longest field goal in NFL history, which he set in 1970 using an extremely deformed foot missing any toes.

Joaquin Phoenix.

Cleft lips or palates also occur relatively frequently at around 1 in 600-800 births, and again can usually be very effectively corrected. Cleft lips appear as a gap or dent in the lip which can continue into the nose, while a cleft palate occurs when the two plates of the skull that form the palate (roof of the mouth) are not completely joined. Both of these are easily treated with surgery, resulting in small scars, which proved little hindrance to the faces of American television and movie stars Thomas John Brokaw, Stacy Keach, Cheech Marin, Jason Robards Jr. and Joaquin Phoenix. Cleft palates have also had little effect on the voices of a number of famous singers born with the condition, such as Richard Hawley of *The Longpigs fame*, Canadian country and folk singer MacNeil and Carmit Bachar, singer with the *Pussycat Dolls*, who has started up an organisation for children born with a cleft lip/palate called *Smile With Me*.

While many inherited diseases present with a shortage of RBCs, leading to anaemia, there are different inherited mutations that can result in increased RBC numbers. One such mutation increases the efficiency of a hormone secreted from the kidneys, called erythropoietin (EPO), to stimulate the bone marrow to produce more RBCs. The increased number of these cells and the corresponding increase in oxygen levels in the blood can lead to greater athleticism. However, this disorder and other disorders leading to increased RBCs (known as polycythemia) can also lead to capillaries becoming clogged by the increased viscosity of the blood. As a result, untreated sufferers are at a risk of thrombosis, heart attacks and strokes. As a treatment for anaemia medical scientists have designed drugs that mimic this erythropoietin hormone in order to stimulate RBC production. It was not long before some sportsmen realised that EPO drugs could help their blood to carry more oxygen and so allow their bodies to work harder for longer. Consequently, by the late 1990s, EPO-abuse had rocketed with athletes seemingly blind to the dangers associated with increased blood hyperviscosity; it has been linked, for example, to the deaths of 18 Dutch and Belgian cyclists between 1987 and 1990. Despite this, allegedly, EPO abuse is still rife in professional cycling.

Blood clotting, such as from a cut on the skin, is accomplished by a complicated process involving many different proteins called coagulation or clotting factors that circulate in the blood. These clotting factors, when activated by an injury result in a blood clot consisting of a plug of platelet cells enmeshed in a network of insoluble fibrin molecules. Deficiencies in some of these blood clotting factors lead to a bleeding disorder known as haemophilia, which affects around one in 5,000-7,000 males. Haemophilia mostly affects males as the genes for the two major affected blood clotting factors, factor VIII (resulting in haemophilia A) and factor IX (haemophilia B), are located on the X chromosome and so are X-linked inherited. This results in affected individuals bleeding for prolonged intervals lasting for days and even weeks in response to only relatively minor injuries. In addition, internal bleeding can also prove fatal.

Haemophilia could be the earliest illness in the world to be recognised as being inherited. Well over 2,000 years ago, Jewish rabbis referred to a bleeding condition that was fatal to young boys and ran in families, so drawing up guidelines to exempt males from circumcision if they had at least two brothers who had previously died from the procedure. However, it was not until the nineteenth century that the illness was properly understood, acquiring the name of haemophilia from a Swiss doctor in 1828.

The British royal family has included many males with haemophilia in addition to female carriers including Queen Victoria herself, as previously described. Victoria seems to have spontaneously acquired the haemophilia mutation *de nova*, as her parents' families are not known to have had the disease. Victoria then passed it on to her son Prince Leopold, Duke of Albany and, through several of her daughters, to various royal houses across the continent such as the Spanish, German and Russian royal families. Prince Leopold, Duke of Albany was diagnosed as a child. Finding that his joint pain, which was another symptom of his haemophilia, was aggravated by cold weather, he would seek warmer climates abroad during the winter. It was while in the south of France, in March 1884, that he slipped and fell in the Yacht Club in Cannes, injuring his knee and dying next morning. His only daughter, Princess Alice of Albany, then passed the gene on to her oldest son, Prince Rupert of Teck, who bled to death after a car accident at the age of 20. Queen Victoria's granddaughter, Czarina Alexandra of Russia, passed the disease on to her son Alexei. Recent genetic analysis of the remains of the Russian royal family have revealed that Alexei had hemophilia B with a mutation in the gene for clotting factor IX.

Only a few years after Alexei was born, the Swedish artist Ivar Arosenius, renowned for his paintings of fairy tales died of the disease at a relatively young age. He depicted his affliction in one of his more famous paintings of "*Saint George Slaying a Dragon*" which can be seen to bleed profusely. The Swedish Haemophilia Society has since established an *Arosenius Fund* to support haemophilia patients.

Even though the illness has affected those at the highest levels of society, it had remained incurable, with an expected lifespan for sufferers of only 20 years. It is only relatively recently, following the development of blood transfusion, that treatment involving the infusion of blood clotting factors has been available.

Alexei Romanov. Circa 1911

However, in the UK, contaminated blood has led to widespread infection of HIV and Hepatitis C with around 1,200 people being infected with HIV and more than 4,800 people infected with Hepatitis C from transfusions. From 1986, heat and chemical treatment of blood products has been used to eliminate such viruses, and since 1992 'recombinant' clotting factor, produced by genetic engineering (see *Endocrine Disorders*) . Unfortunately, there is still no permanent way of replacing or enabeling the body to synthesise a functional clotting factor.

In addition to blood clotting factors, platelet cells also play an important role in controlling bleeding and inflammatory response. Low platelet levels or dysfunctional platelets (such as in thrombocytopenia and Gaucher's disease) predispose to bleeding, while high numbers of platelets can lead to thrombosis. When blood vessels are damaged, collagen becomes exposed, to which the platelets aggregate, binding to each other via a receptor they contain on their cell surface known as glycoprotein IIb/IIIa. Defects in this receptor can lead to reduced platelet adhesion and prolonged bleeding times, a condition known as Glanzmann's thrombasthenia. Another very common mutation in a gene for the receptor, carried by about 20% of the population, can predispose individuals to a higher risk of developing acute coronary artery events. The sudden death of the Russian Olympic Gold medallist skater Sergei Grinkov has been linked to this phenomenon. On November 20th 1995, at the age of 28, Grinkov died suddenly of a massive heart attack, while in the middle of a practice session. Grinkov's father also died at the age of 52 of a heart attack, and like his son had none of the risk factors associated with heart disease such as smoking, diabetes, high blood pressure and elevated cholesterol levels. Yet an autopsy revealed that he had severe coronary artery disease and subsequent DNA tests showed that he carried the gene variation

Saint George Slaying the Dragon. Ivar Arosenius, 1908.

Another protein involved in controlling platelet activation is the von Willebrand factor (vWF) protein. Mutations in the gene which codes for this protein lead to the most common hereditary coagulation disorder, von Willebrand's disease. Three different types of von Willebrand's disease exist differing in severity, from the rare severe form to the relatively common mild forms characterised by only slightly prolonged bleeding, possibly affecting as many as 1% of the population. These mild forms can be a complication in child abuse cases, because it causes children to bruise easily. One person with the severe type of von Willebrand disease is the musician Bill Kozlowski, who was founder of the Alaskan rock band *Peabody's Monster*. Much of Kozlowski's song-writing focuses on living life to the fullest as he knew his lifespan would be limited. Since his death in 2003, many of his compositions have been made available on his memorial site, billkozlowski.com.

Stamp commemorating Ekaterina Gordeeva and Sergei Grinkov, 1998

In contrast to bleeding disorders, thrombotic diseases are characterized by the formation of a thrombus, or blood clot, which can obstruct blood flow. It is suggested that, of all the patients suffering from thromboembolisms, 25-50 percent have a genetic predisposition. This results from defects in genes producing proteins inhibiting blood clot formation as part of the feedback pathway shutting off blood coagulation once a clot is already formed. One anticoagulant prescribed to people with increased tendencies for thrombosis is warfarin, which inhibits the production of some of these clotting factors. Initially it was developed as a rat poison, but it was not until a man in the US Navy tried, unsuccessfully, to commit suicide in 1951 by drinking it, that its use as a therapeutic anticoagulant was considered.

Chapter 10

IMMUNE DISORDERS

It is a fair, even-handed, noble adjustment of things, that while there is infection in disease and sorrow, there is nothing in the world so irresistibly contagious as laughter and good-humour.

Charles Dickens

When one or more components of the immune system are defective immunodeficiencies occur. Primary immunodeficiencies result from genetic defects and usually present with recurrent or overwhelming infections in very young children. Secondary immunodeficiencies, on the other hand, are acquired in various ways during life, perhaps by malnutrition or else by an infection of some kind such as HIV. The immune system basically describes the collection of biological processes within an organism that serve to protect it against disease by identifying and killing pathogens. Vertebrates, such as ourselves, have evolved a special type of immune response known as the adaptive immune response which basically allows the immune system to recognize and remember specific pathogens (to generate immunity), and to mount stronger attacks each time the pathogen is encountered. Phagocyte cells in the blood are able to engulf antigens (foreign cells/proteins) and display them on their cell surface. Other cells found in the blood, known as T-lymphocytes (T-cells) then recognise these foreign proteins and release chemicals to further break down this foreign material and also release additional molecules that attract and stimulate the division of further immune cells. These antigens also bind to the cell surface of another type of immune cell, B-lymphocytes (B-cells), causing these cells to divide and produce antibodies (immunoglobulins; Ig). Antibodies then function to destroy antigens by causing them to precipitate or clump, and can in addition initiate further pathways that cause the destruction of the antigen and stop the antigen from entering host cells. Finally, some of the activated B cells and T cells then go on to become memory cells, dividing and reproducing the same antigen through the individual's lifetime. These B and T lymphocytes form part of the immune response, as these memory cells are able to respond extremely quickly to an antigen if it is recognised from the previous infection.

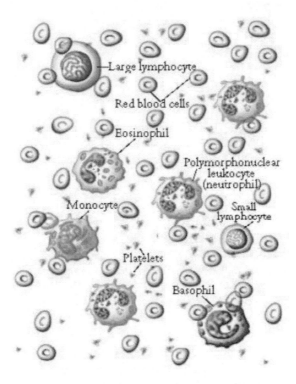

Different types cells involved in the immune system.

Perhaps the earliest documented incidence of acquired immunity was during the plague of Athens in 430 BC. Here it was noted at the time that people who had recovered from a previous bout of the disease could nurse the sick without contracting the illness a second time. This forms the basis of vaccination whereby a substance is introduced into the body in order to produce immunity to a disease. Although it may have been used in China and India in ancient times, the first recorded vaccination (Lat. *vacca*; cow), was by Edward Jenner in 1796 when he used cowpox to give immunity to smallpox after observing that dairymaids who had previously been sick with cowpox (a contagious disease that causes blisters on the cow's udder and on the milkmaid's hand) did not catch smallpox. Pasteur then further developed the use of artificial vaccines that could be used for a variety of diseases such as rabies by growing the virus in dogs and rabbits, then weakening it by drying the affected tissues, and finally reintroducing it into healthy animals. This serves to activate an immune response and form an immune memory so that when the animal comes in contact with the infectious form of the virus its immune system can quickly respond.

Before the advent of antibiotic drugs most babies born with inherited immune defects would have died early due to their susceptibility to infections and such cases would not have been easy to identify among the many normal infants who would have also died of infections. It was not until 1952 that the first immunodeficiency disease was described, though many inherited immunodeficiencies are now known. These are classified according to which component of the immune system is deficient. The most common inherited immunodeficiency is selective IgA antibody deficiency, which may be present in 1 in 400 people. While it seems to result in no specific symptoms it is found more often in people with chronic lung disease. Apart from IgA deficiency, the overall incidence of primary immunodeficiency is around 1 in 10,000.

The first person to describe an inherent immunodeficiency disease was Dr. Colonel Ogden C. Bruton in 1952, when he reported the failure of a male child to produce antibodies due to a defective gene hindering B-cell growth. This is now known as Bruton's X-linked agammaglobulinemia, as the gene involved locates to the X chromosome. Dr Bruton also became the first physician to provide specific immunotherapy for this disorder by administering injections of antibodies to the patient who responded well to the treatment. Since then, many more diseases of antibody production have been described, where affected infants usually develop recurrent infections with pyogenic, or pus-forming, bacteria such as pneumonia-causing *Streptococcus pneumoniae*.

While patients with B-cell defects can still deal with many pathogens adequately, patients with defects in T-cell development are highly susceptible to a broad range of infectious agents, reflecting the important role T-cells have in many parts of the immune system. Such diseases are known as Severe Combined ImmunoDeficiency (SCID) as patients, unable to generate any immunological memory or cell-mediated immune responses, suffer from infections from many opportunistic pathogens. This results in early death if untreated with either bone marrow transplantations or enzyme replacement therapy. SCIDs have an estimated incidence of 1 in 100,000, and the most common type of SCID is called XSCID due to the mutated gene being found on the X chromosome. This gene normally codes for part of a receptor on the surface of lymphocytes allowing them to mature, proliferate and mobilize to fight infections.

David Vetter.

Awareness of these diseases was raised in the 1970s by the plight of David Vetter, who lived nearly all of his 12 years of life inside a sealed plastic bubble designed to protect him from infections. David Vetter's parents first had a daughter, Katherine, who was normal and then a son who died six months after birth from SCID. They realised that any subsequent child would have a 50 percent chance of inheriting the same condition but believing that a bone marrow transplant from their healthy daughter would give any infected child a normal healthy life, they decided to go through with another pregnancy. Twenty seconds after being removed from his mother's womb, in September 1971, David Vetter was sealed in a germ-free environment that would be his home for nearly all of the rest of his life. Water, air, food, diapers and clothes were all disinfected with special cleaning agents before entering his cocoon and he was handled only through special plastic gloves attached to the walls. Eventually, a bone marrow transplant was performed using marrow donated by his sister Katherine. However, a few months after the operation, David started having diarrhoea, fever and severe vomiting, requiring him to be taken out of the bubble for the first time in 1984, for treatment. Sadly though, he died 15 days later of a lymphoma, caused by an unscreened virus, in his newly transplanted bone marrow.

GENE THERAPY

Gene therapy is the insertion of genes into an individual's cells and tissues to treat genetic diseases by supplementing a defective mutant allele with a functional one.

In the 1980s, methods became available to allow easy ways to produce proteins, such as the insulin protein, deficient in diabetics, or clotting factors missing in patients with haemophilia. This is done by introducing human genes into bacterial DNA, so that the modified bacteria can then produce the corresponding protein, known as a recombinant protein, which can be harvested from the bacteria and injected in people who cannot produce it naturally. Scientists took the logical step of trying to introduce genes straight into human cells, focusing on diseases caused by single-gene defects, such as cystic fibrosis, haemophilia, muscular dystrophy and sickle cell anaemia.

Target cells such as the patient's liver or lung cells are infected with a viral vector, used to carry genetic material containing the therapeutic human gene, into the target cell. The generation of a functional protein product from the therapeutic gene restores the target cell to a normal state.

In theory it is possible to transform either somatic cells (most cells of the body) or cells of the germline (such as sperm cells, ova, and their stem cell precursors). All gene therapy so far in people has been directed at somatic cells, whereas germline engineering in humans remains only a highly controversial prospect since the introduced gene could be transmitted to offspring. Somatic gene therapy can be broadly split in to two categories: *ex vivo* (where cells are modified outside the body and then transplanted back in again) and *in vivo* (where genes are changed in cells still in the body).

On September 14, 1990 researchers in the U.S. performed the first approved gene therapy procedure on four-year old Ashanti DeSilva who was born with SCID. The procedure involved removing white blood cells from her body, inserting the missing gene into the cells, and then infusing them back into her bloodstream. Tests showed that the therapy strengthened Ashanti's immune system though she needed to continually receive replacement gene therapy. Therefore, while this procedure is not a cure it did prove that in theory the symptoms could be alleviated. However trials in France highlighted the possible pitfalls to such a procedure with a number of children treated for XSCID with gene therapy, developing leukaemia due to the randomly inserted gene sometime disrupting further genes leading to cancer. As a result of this, gene therapy treatments for SCID were discontinued for a long period of time. However newer approaches combined with using different modified viruses to introduce genes into cells are beginning to reopen the door to treating a range of diseases from haemophilia, adrenoleukodystrophy and some forms of blindness.

Phagocytic (Gr. *phage*; eat) cells destroy foreign matter, such as microorganisms by ingesting them into their cells where various enzymes reactive oxygen-containing molecules are present to break down the pathogen. There are a number of gene mutations that can lead to defects in the ability of phagocytes cells to kill pathogens. This occurs in chronic granulomatous disease due to mutations in a number of different genes. Affecting around 1 in 200,000 people, the disease results in recurrent chronic bacterial infections. Another part of the immune system is composed of a pathway of different proteins, known as the complement system. This cascade of many small plasma proteins, when activated, can work in many ways to actively destroy pathogens or label them for destruction by phagocyte cells. The first hereditary deficiency of a complement protein was actually discovered in immunologists using their own blood for experiments reflecting the fact that some parts of the complement system do appear to have some redundancy. However, the absence of some specific proteins involved in the complement can lead to highly specific bacterial infections; one such gene defect leads to a 10,000 fold increased risk of *menigococcus* infections and consequent meningitis.

While deficiencies in the immune system can lead to infections, an overactive immune system can lead to another class of diseases known as autoimmune diseases. This is characterised by a decreased ability of the immune system to distinguish foreign material from host cells resulting in the immune system attacking the individual's own tissues. There are more than forty different autoimmune diseases affecting 5% to 7% of the population. Women account for a greater proportion of the sufferers - nearly 80% in the USA - which may be due to the role of sex hormones.

Some people are born with an inherited genetic susceptibility to autoimmune diseases. Such genetic risk factors often involve the inheritance of specific variants of genes called Human Lymphocyte Antigens (HLA). The function of these HLA proteins, displayed by all cells of the body, is to present fragments of antigens, either from the host cell or from invading pathogens on the outside of the cell for passing T-cells to recognise. In this way they can be thought of as acting as a sort of signpost in allowing the immune system to differentiate between host and foreign cells. Humans make at least 100,000 proteins that can be processed into 3 million different smaller fragments for presentation to T-cells. Problems can occur though when some HLA alleles present foreign materials to the T-cells that happen to resemble proteins found in the body, so stimulating the immune system to react against these proteins that may be found on completely healthy cells. In addition, other HLA alleles exist which are less efficient at presenting proteins found in the host body. This can lead to the development of immune cells that are unable to recognise certain proteins found naturally in the body and consequently form an immune response when coming into contact with these proteins.

Nadezhda Krupskaya.

Depending upon which tissue in the body the immune system attacks, a number of different diseases can develop. For example, the production of antibodies against tissues of the thyroid gland can lead to over manufacture of thyroid hormone resulting in hyperthyroidism (an overactive thyroid) seen in Grave's disease. President George H. W. Bush was diagnosed in 1991 shortly after his wife Barbara and their dog *Millie*. Because of the remarkable coincidence of all three cases of auto-immune disease in one household, it was suspected that this may have been the result of a plot to poison the President and the Secret Service went as far as to test the water in all of the presidential residences, including the White House, for any toxins. Many pictures of Barbara Bush during this period show that she had developed many of the classical characteristics of the disease, such as a bulging of the eyes and a throat goitre. It is thought that the wife of Vladimir Lenin, Nadezhda Krupskaya, also suffered from Graves's disease with her bulging eyes giving her the Bolshevik codename of *Fish*. Furthermore, since Graves Disease can also disrupt the menstrual cycle, it is believed that this is why the couple never had children.

Richard Pryor.

The body may also start to produce an antibody against specific cell receptors, such as neurotransmitters in the disease Myasthenia Gravis. This has the effect of blocking nerve signals to muscle cells resulting in muscle weakness. Suzanne Rogers, an American actress best known for her role in the daytime television series *"Days of Our Lives"*, was diagnosed with Myasthenia Gravis in 1984, and insisted that her character, *Maggie*, be diagnosed with the same disease on the show in order to raise awareness of the condition..

When the body's immune system attacks the myelin sheath surrounding nerve cell axons, multiple sclerosis (MS) occurs. Translating as "multiple hardenings", the most popular hypothesis as to the cause of MS is that it results from an infection which primes a susceptible immune system to destroy myelin later in life. The body then tries to repair the damage with hard plaques of scar tissue which further disrupts the flow of electrical impulses. Although not considered a hereditary disease, genetics does play some role in determining susceptibility to MS, particularly the inheritance of certain HLA variants. The contribution of genes to MS is evident from studies in identical twins where the likelihood that the second twin may develop MS if the first twin does is about 30%. Moreover, there are some populations, such as the Inuit and the Japanese who have very low incidences. There are several historical accounts of people who suffered from MS. One of the first identifiable MS patients was Saint Lidwina of Schiedam (1380–1433), a Dutch nun, who from the age of sixteen until her death at age 53, suffered from intermittent pain, weakness of the legs, and vision loss. The earliest person to have a direct diagnoses of MS was Sir Augustus Frederick d'Este, grandson of George III, who left a highly detailed diary describing his 22 years living with the disease. His symptoms began at the age of 28 when he began to develop weakness of the legs, clumsiness of the hands, numbness, dizziness and bladder disturbances that worsened until 1844, when he was confined to a wheelchair. The American comedian, Richard Pryor, in his later years became a wheelchair user due to MS, which he said stood for *More Shit*.

Seal.

While some autoimmune diseases are due to an immune response to a specific tissue, others are characterised by general tissue damage throughout the body, i.e. systemic. This is seen in systemic lupus erythematosis where antibodies are produced to cells in many tissues of the body. The term lupus is attributed to the 12th century physician Rogerius, who used it to describe the classic reddish, butterfly-shaped rash across the nose and cheeks - the name either deriving from the Latin for wolf as it can resemble the pattern of fur on a wolf's face, or from a French style of mask which women reportedly wore to conceal the rash on their faces. The English singer/songwriter, Seal, was diagnosed with the skin-affecting variant of this condition as a young man which has left him with the distinctive facial scars on his cheeks. Systemic lupus erythematosis can cause the immune system to attack and damage various parts of the body, including the joints, skin, kidneys, heart, lungs, blood vessels, and brain, but symptoms vary enormously in type and severity. While research indicates that systemic lupus erythematosis has a genetic link and can run in families, no single gene has yet been identified, as is the case with the majority of autoimmune diseases.

Chapter 11

ENDOCRINE DISORDERS

War will never cease until babies begin to come into the world with larger cerebrums and smaller adrenal glands.

Henry Louis Mencken (US Journalist).

The endocrine system of the body comprises a group of glands and organs that regulate and control various body functions by releasing specific proteins or steroids known as hormones. These are released into the bloodstream where they act as messengers, affecting the various activities of different parts of the body. When a particular hormone reaches its target cell it transmits its message by binding to a receptor, on the cell surface or in the cell nucleus, causing the cell to take a specific action. In this way, very small amounts of hormones can trigger very large responses in the body, controlling the function of entire organs, affecting such diverse processes as growth and development, reproduction, and sexual characteristics.

Although hormones circulate throughout the body, each type of hormone influences only certain organs and tissues. Some hormones affect only one or two organs, whereas others have influence throughout the body. The endocrine system comprises the hypothalamus, the pituitary gland, the thyroid gland, the parathyroid glands, the islets of the pancreas, the adrenal glands, the testes in men, and the ovaries in women. The secretion of each hormone must be regulated within precise limits, and so many endocrine glands are controlled by the interplay of hormonal signals between the hypothalamus, located in the brain, and the pituitary gland, which sits at the base of the brain.

The hypothalamus is a region of the brain that secretes a number of hormones known as releasing hormones, as these further stimulate the secretion of hormones from the pituitary gland. For example, gonadotropin-releasing hormone released from the hypothalamus signals the pituitary to produce follicle stimulating hormone and leutenising hormone which, in turn, stimulates the ovaries or testes to produce yet further hormones, such as testosterone and oestrogen, to signal sexual development. An inherited deficiency of gonadotropin-releasing hormone, therefore, results in the absence or decreased function of testes or ovaries, and is known as Kallmann syndrome. First described in 1944 by the German geneticist, Franz Josef Kallmann, and occurring at a rate of 1 in 10,000 male births and 1 in 50,000 female births, this syndrome generally presents with hypogonadism (i.e. an underproduction of testosterone), delayed puberty and a lack of secondary sex characteristics, such as breast development.

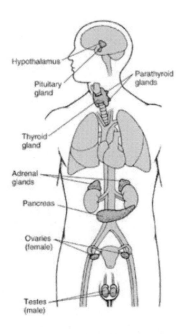

The endocrine system.

Two further hypothalamic hormones are vasopressin and oxytocin which are released through the pituitary. A major function of vasopressin is to maintain water balance by regulating levels of water excreted by the kidneys. A lack of vasopressin leads to large amounts of water being excreted by the kidneys; a condition known as central diabetes insipidus. Deriving from the Greek verb *diabainein,* which means to stand with legs apart, a person with diabetes insipidus can urinate as much as 18 litres daily - the average urine volume for a normal adult is 1.5 litres - which is naturally also accompanied by excessive thirst. The Latin word insipidus means without taste, referring to the diluted nature of the urine. This is in contrast to mellitus (as in diabetes mellitus) describing the excretion of sweet urine in an entirely separate endocrine disease where defects in insulin hormone lead to high blood glucose levels. A synthetic drug that mimics the action of vasopressin, known as Desmopressin, is used to reduce urine production in central diabetes insipidus, and may also be prescribed to reduce frequent bedwetting episodes in children by decreasing nocturnal urine production. Though diabetes insipidus must have been evident in a great many number of individuals through the ages, it was not until 1792 that this genetic disorder was eventually described in the medical literature. This detailed the story of a French woman who drank over 20 litres of water a day. Since the age of 3 she had remembered drinking buckets of water, until, in her early teens, she was forced out of home by her uncompassionate, and probably bewildered, parents. This disease is usually caused by mutations in the gene for vasopressin that regulates levels of water excreted by the kidneys and affects around 1 in 30,000 individuals. A far rarer form of diabetes insipidus results from mutations in a receptor for vasopressin found on the X chromosome and so leading to sex-linked inherited diabetes insipidus, specifically affecting males. This was long ago described in Scottish folklore where, according to legend, a gypsy woman travelling with her thirsty son was denied water by a housewife. The gypsy woman then cursed the housewife, causing the housewife's sons to crave water while condemning her daughters to pass the curse on to future generations. And it appears that they did pass on the trait with two Scotsmen immigrating with the gene to Nova Scotia, Canada, in 1761 arriving on a ship named the *Hopewell*. As a result, this disease is relatively common in some regions of Canada. Considering that fresh water supplies on board ship would have been limited and carefully rationed, the journey for those two Scottish immigrants must have been a considerable ordeal.

The pituitary (Lat. *pituita*; mucus, as this gland was once thought to produce nasal mucus) gland is a pea-sized structure located at the base of the brain. Known as the master gland, it secretes several hormones including growth hormone which regulates growth thyroid-stimulating hormone which stimulates the thyroid gland to produce thyroid hormones and adrenocorticotropic hormone, which stimulates the adrenal glands to produce cortisol. It also produces follicle-stimulating hormone, luteinizing hormone, prolactin, melanocyte-stimulating hormone and endorphins that inhibit pain sensations and help control the immune system.

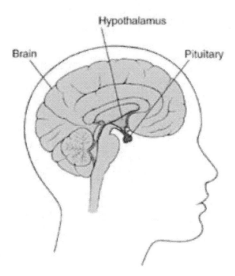

Pituitary and hypothalamus.

The pituitary gland can malfunction in several ways, usually as a result of developing a noncancerous (benign) pituitary tumour that may lead either to an overproduction or an underproduction of one or more hormones. Often remaining undiagnosed these tumours are found in around 15% of the population. While most such pituitary tumours are generally sporadic, some do have a genetic basis as mutations in some genes can lead to a hereditary predisposition. Depending on the size and nature of the tumour, too little or too much of a particular pituitary hormone can lead to a wide variety of symptoms.

The pituitary constantly produces a regulated supply of growth hormone that, secreted into the bloodstream, stimulates cells of the body to grow and divide ensuring that all portions of the body grow at an equal rate. Increased production of growth hormone can occur as a result of a pituitary tumour. When this tumour develops during childhood a condition known as gigantism can occur characterised by lengthening bones, especially in the arms and legs, resulting in the individual growing to an unusually great height. Generally, a clinical definition of a giant is someone standing over 7ft 6in tall. Perhaps the first written account of a true giant comes from the bible, as Goliath was said to stand *"taller than six cubits"* which equals about nine and a half feet. This may not be so farfetched because people afflicted with gigantism continue to grow throughout their lives. One well-known individual with gigantism, called Charles Byrne, had an enlarged pituitary comparable in size to a peach. Growing to a height of 7ft 9in, he was constantly hounded by doctors desperate to get his body on their autopsy table. One particularly assiduous doctor, John Hunter, was particularly keen to boil his corpse the moment the poor man died in order to examine the skeleton. It was supposedly this that drove the poor man into drinking himself to death in 1783 at the age of 22. So afraid was he that John Hunter and the other doctors would get their hands on his body, Charles Byrne requested from his deathbed that he be buried at sea. However, John Hunter was quickly on the scene to bribe the hospital staff, purchasing the cadaver for 500 pounds. His skeleton still resides in the Royal College of Surgeons.

Charles Byrne.

When too much growth hormone suddenly begins to be produced in adulthood, usually due to the development of a pituitary tumour, long after the growth plates of the bones have closed, a disorder referred to as acromegaly (Gr. *acro*; end, *mega*; large) occurs. This results in bones becoming deformed rather than elongated. Often, the jawbone can overgrow and cartilage in the voice box can thicken, making the voice deep and husky. A major complication is joint pain and arthritis with some individuals also feeling weaknesses in their limbs, loss of vision and severe headaches as the enlarging tissues compress nerves. Film actors with this condition have inevitably been cast in intimidating roles such as Richard Kiel who played *Jaws* in the *"James Bond"* movies, and also Ted Cassidy and Carel Stryken who both played the character *Lurch* in *"The Addams Family"*. Rondo Hatton who played a number of thuggish parts during his movie career serves as an extreme example of how the disease can distort the skull and face.

Rondo Hatton.

If the pituitary gland does not produce enough growth hormone in childhood, then an abnormally short stature may result, affecting all body parts to an equal degree. Often referred to as proportional short-stature, such individuals can often physically resemble a child. The memoirs and life story of Count Josef Boruwlaski, who suffered from such a growth hormone deficit, reveals a great deal about the condition. Born in Russia in 1739 to parents of normal size, he grew no higher than 71cm. At the age of nine a wealthy noblewoman, Staorina de Caorliz, took a shine to Josef Boruwlaski and he went to live with her to be educated. As a result, although he only stood two feet tall in his early teens, he possessed perfect etiquette and was a brilliant composer of music, and as a result was accepted into the highest levels of society. One of Josef Boruwlaski's four siblings also showed severe height restriction, while the other three grew to a normal stature, suggesting that as yet unidentified genetic mutations may well be behind the condition. It is thought that similar mutations affect pygmy tribes (a term which is sometimes applied to communities in which the height of individuals does not exceed 150 cm). One such mutation found in the gene for growth hormone receptor occurs in a group of people living in the foothills of the Ecuadorian Andes. Interestingly, recent research shows this gene may not be indigenous as exactly the same gene mutation has been identified in a number of Israelis suggesting that the mutation may have been carried to the region by Spanish Jews immigrating in the early 1500s to escape persecution.

Another hormone secreted from the pituitary is prolactin. Increased concentrations of this hormone during pregnancy cause enlargement of the mammary glands and an increase in the production of milk. However, around 30% of pituitary tumours lead to increased prolactin production and increased lactation in women who are not pregnant. Accompanying the increased lactation is suppression of ovulation (prolactin inhibits ovulation to stop further pregnancies occurring during breastfeeding). Often women with this tumour visit their gynaecologist thinking that they might be pregnant. This may have affected Queen Mary I of England, who died childless despite having symptoms suggestive of pregnancy on several occasions, including swelling breasts and discharges of milk.

African pygmies and Professor K. G. Murphy.

The thyroid gland is situated to front of the neck below the pituitary gland.. It is named after the thyroid cartilage; a *thyreos* was an ancient Greek army shield shaped like a door, with a notch at the top for the soldier's chin. The thyroid cartilage is also referred to as the Adam's apple and serves to shield the vocal cords of the larynx. The thyroid cartilage is usually about the same size in both girls and boys, but after puberty testosterone causes the larynx to grow in males much more than in females. This is the reason why voices break in teenage boys.

The thyroid gland secretes the hormone thyroxine that controls metabolism, respiration, and heart function. Mutations disrupting the function of thyroxine result in a disorder known as cretinism, characterised by severely stunted physical and mental growth. However, the majority of cases of cretinism occur due to a lack of iodine in the diet during childhood. This is because the production of thyroid hormones requires iodine. *Crétin* is an old French name for the wild men of the Alps in French folktales as this disease was for many centuries more common in areas where the soil is deficient in iodine – particularly in the Alpine regions of Europe. A survey in an area of Switzerland in 1810 revealed that among 70,000 inhabitants, 4,000 individuals suffered from cretinism. A further sign of this disease is enlarged thyroids, known as goitres, as the thyroid attempts to make up for the hormone reduction. This used to be common in the Carboniferous Limestone area of Derbyshire where soils are naturally low in iodine, and for this reason the condition was often referred to at the time as "Derbyshire Neck". This problem was eventually tackled with the introduction of iodised milk, generally produced by increasing the iodine content of animal feed.

Queen Mary I of England. Anthonis Mor, 1554

An underactive thyroid gland resulting in thyroid hormone deficiency, when present at birth, is known as congenital hypothyroidism and affects approximately 1 in 4,000 newborn infants. Regarded as being the most prevalent inborn endocrine disorder it can result from mutations in a number of different genes involved in thyroid hormone synthesis. If not treated with daily doses of thyroid hormone, growth failure and permanent mental retardation can develop. In contrast, the thyroid gland can also become overactive, known as hyperthyroidism, resulting in the release of excess thyroid hormones. This occurs in a number of diseases such as, the previously mentioned Graves' disease, resulting in weight loss (often accompanied by a ravenous appetite), fatigue, and a number of other symptoms.

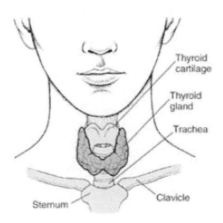

Thyroid gland.

The Olympic skier Janica Kostelic suffers from Graves' disease. She first noticed something was wrong on a glacier in Switzerland, when she collapsed with breathing problems and had to be airlifted to hospital. After being diagnosed with hyperthyroidism, she had her thyroid removed at the beginning of 2005. She went on to win gold at the 2006

Winter Olympic Games in Turin - her fourth Olympic Gold medal - making her the most successful female skier in the history of the Olympic Games.

The pair of adrenal (Lat. *ad;* above, *renes*; kidneys) glands are located just above the kidneys. These secrete two steroids, aldosterone, which acts to conserve sodium ions and water in the body, and cortisol which increases blood glucose levels. They also secrete two hormones, epinephrine and norepinephrine, particularly during stress. When the adrenal glands become underactive, Addison's disease occurs. First described in 1855 by Thomas Addison, who committed suicide when his work was largely ignored and refused publication, this disease may be due to a genetic abnormality or else may result from an autoimmune reaction, whereby the body's immune system attacks and destroys the adrenal tissue. A deficiency of aldosterone disrupts the kidneys ability to regulate water and salt balance leading to reuced blood pressure. In addition the inability to produce corticosteroids can lead to an extreme sensitivity to insulin so that the level of sugar in the blood may fall dangerously low. Weakness of muscles and the heart may also occur, and as a result the body may be unable to react properly under stress leading to possible severe medical complications. The pituitary gland, in an attempt to stimulate the adrenal glands, becomes more active leading to increased melanin production and darkened skin.

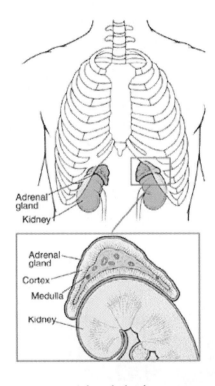

Adrenal glands.

President John F. Kennedy was another well cited sufferer of Addison's disease and the suspected occurrence of the same disease in his sister, Eunice Kennedy Shriver, may argue for an inherited cause of his disease. At the age of 30, whilst in the UK, he was so ill that a doctor gave him less than a year to live and he was even given the last rites by the ship's chaplain on his sea voyage home. Only 13 years later, when he ran for president, he appeared fit, healthy and tanned, and this no doubt influenced the voters. Ironically, his tan may well have been a side effect of his disease. The heavy amounts of steroids he had taken since childhood for his condition also resulted in osteoporosis and back pain which required him in addition to take cocktails of very strong pain medications. Many photographs from this time seem to show the swelling effects of the steroids, particularly in the face. The steroids could also have increased his libido, which may add weight to the many rumours surrounding JFK's supposedly insatiable sex drive. Another person thought to have suffered from Addison's disease was Jane Austen who died in 1817 at the age of 42. Her last words of "*Nothing, but death*", when asked by her sister Cassandra if there was anything she wanted, highlights the misery the disease could cause before the advent of modern-day treatment.

President John F. Kennedy, 1961.

In 1973, a white 40 year old South African housewife, by the name of Rita Hoefling, had an operation to remove her adrenal glands. This led to the increased production of the hormone melanotropin that stimulates melanocyte cells in the skin to increase the production of pigment, as previously mentioned. As a result she started to turn a deep bronze colour. The pro-apartheid Afrikaners were not the most tolerant of people at this time. Refused entry to white only areas and rejected by her family and friends, she was obliged to move to a non-white area of Cape Town. After five years or so, her skin pigment returned to normal but by then there could be no going back to her old life, and she eventually she died alone in a bedsit at the age of 55.

The reason Rita Hoefling had her adrenal glands removed, was because she had been diagnosed with another adrenal disorder known as Cushing's syndrome. This condition occurs when the adrenal glands become overactive, often due to a benign tumour, producing high levels of steroids, which in turn control the amount and distribution of body fat. As a consequence, this syndrome results in excessive fat developing - a typical sufferer usually has a large, round face with arms and legs that are usually slender in proportion to the thickened trunk. In addition, these steroids also raise levels of testosterone which can lead to increased facial and body hair in women and balding in men. It has been suggested that Henry VIII, the notably un-merry monarch, suffered from Cushing's syndrome, which would have accounted for the bloated face seen in his portraits and the massive bulk of his body. The fact that this disorder can also cause irritability, depression, aggression, psychosis and impotence, might explain a few other things as well.

Portrait of Henry VIII. Hans Holbein the Younger, 1937.

Women are never disarmed by compliments. Men always are.
That is the difference between the two sexes.

Oscar Wilde

The testes have been regarded as the source of maleness since ancient times. The way in which the testes control maleness is by producing male sex steroid hormones known as androgens (including testosterone) that stimulate development of the male sex organs in addition to other male characteristics. The ancient Greek philosopher Hippocrates was one of the first to note this, describing how castrated male children grew up to have underdeveloped external genitalia, high-pitched voices, and very little facial or body hair.

Joan of Arc. Circa 1845.

Some genetic males are born with disrupted androgen function, resulting from mutations in the androgen receptor gene located on the X chromosome. This results in a male embryo that, though producing masculinising androgens from embryonic testes, is unable to react to the testosterone. This leads to a hermaphrodite appearance, primarily characterised by female characteristics but with testes that may be more or less evident. Referred to as androgen insensitivity syndrome (AIS), the condition can be so subtle that it may go unnoticed until after puberty; most individuals with the condition do not menstruate and hence are rendered infertile. Affecting approximately 1 in 20,000 individuals, people with AIS also tend to be taller, as the male Y chromosome is mainly responsible for growth, and slimmer due to the affect of the small amounts of oestrogen produced by the testes that in addition can stimulate generous breast growth. Also, due to low levels of androgens, girls with AIS will generally not suffer from acne and often present with beautifully clear skin, luxuriantly thick scalp hair and little or no pubic or underarm hair. The overall effect is that AIS women are often exceptionally beautiful and, as such, sufferers are often found in occupations such as modelling or acting.

A number of well known women in history may well have suffered from AIS. It has been suggested that Queen Elizabeth I, known as the Virgin Queen, may have had the condition, but evidence is slim and seems mainly to derive from the fact that she never married and avoided sexual relations, this raised speculations that did not menstruate. Joan of Arc may well also have had the disorder and even during her life her gender raised speculation. An investigation following her execution in 1431, which received testimonies from over 100 witness statements, established that she had had well developed breasts, no pubic hair, and had never menstruated.

The androgen receptor, as well as binding testosterone, also binds dihydrotestosterone (DHT), a more potent androgen steroid converted from testosterone by the enzyme 5-alpha-reductase. It is DHT that is responsible for early male sex characteristics, with testosterone influencing masculinisation during adolescence. Mutations in the gene for this enzyme in male embryos can lead to a failure of genital growth and development of female characteristics. However, during adolescence, when testosterone takes over, masculinisation occurs with testes descending and the growth of facial hair. In the Dominican Republic the condition is referred to colloquially as "*testes-at-twelve*". In more remote regions of the country, as many as one percent of children regarded as female have to have their gender reassigned at puberty. All these individuals can trace their ancestry to a single woman, by the name of Altagracia Carrasco, who carried the mutation and lived in the mid nineteenth century. One well-known case of this syndrome was a French girl called Herculine Barbin who was born in 1838 in Saint-Jean-d'Angély, France, and raised as a girl. She attended a convent school for girls and began a long affair with the headmistress' daughter, Sara. As she moved into adolescence her breasts failed to develop, facial hair started to appear and she suffered painful abdominal pains. When the doctor was called it transpired that Herculine had descending testes. Reassigning her sex to male and giving him the new name of Abel he was cast out from the school where he went to Paris to work on the railroads. Tragically, a few years later, alone and heartbroken he committed suicide, leaving a book of his memoirs which have since formed the basis for the novel *Middlesex* by Jeffrey Eugenides. Herculine Barbin's birthday, the 8th November, is now designated as the Intersex Solidarity Day.

Conversion of testosterone into estradiol (the major oestrogen in humans) and dihydrotestosterone by the two enzymes aromatase and 5-α-reductase.

Another enzyme called aromatase is responsible for turning testosterone into oestrogen. Testosterone is produced by the mother's placenta during pregnancy, and needs the aromatase enzyme to convert this to oestrogen. Defects in the gene for this enzyme in the mother cause developing foetuses to be exposed to increased levels of testosterone and as a result any girls born may be masculinised. This mutant gene is found in all spotted hyenas, confirming what Aristotle first noted more than 2,000 years ago, that the females of the species are bigger, and more aggressive. Packs of spotted hyenas are led by a single alpha female, whilst the highest ranked male in the group is still subordinate to the lowest ranked female. This is illustrated in the Disney movie "*The Lion King*". The script writers and artists spent some time observing spotted hyenas before creating *Shenzi* the female spotted hyena, as the biggest and strongest in the pack, assuming the role of leader. There is a similar condition that can occur due to the lack of another enzyme, 21-hydroxylase, used in the production of specific steroids, ultimately leading to increased levels of testosterone. Known as congenital adrenal hyperplasia, boys with this disorder enter puberty early, sometimes at 2-3 years of age, developing a deep voice and pubic hair. Females inheriting such a mutation, on the other hand, can develop both female and male genitalia, to an extent that it can be difficult to determine the sex of a newborn. In 1865 an Italian physician, by the name of Luigi De Crecchio, possibly described the first known case of this. "..., *there arrived toward the end of January a cadaver which in life was the body of a certain Joseph Marzo... decidedly male in all respects. There were no feminine curves to the body. There was a heavy beard. ..the lower extremities were somewhat delicate, resembling the female... there was a penis... and there were two folds of skin coming from the top of the penis and encircling it on either side, resembling labia majora*".

Chapter 12

METABOLIC DISORDERS

What is food to one man may be fierce poison to others.

Lucretius (99 BC - 55 BC)

There are many different diseases in which individuals lack a particular enzyme involved in converting various substances into other products, i.e. metabolism. Termed "inborn error of metabolism" by the British physician, Sir Archibald Garrod, in the early 1900s, the diseases highlight the fact that one gene produces one enzyme protein. In other words a mutation in one particular gene will only affect the one particular protein produced by that gene. While seemingly obvious now, this was a very astute observation at the time and helped pave the way for understanding the basis of genes and inheritance.

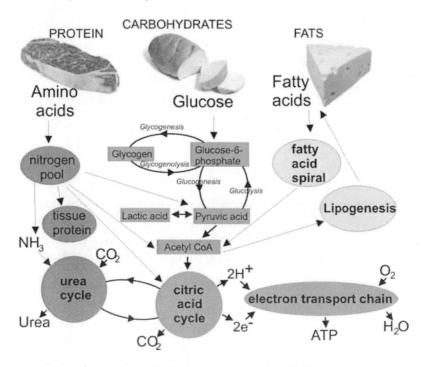

A few of the metabolic pathways in a cell. Energy is conserved in ATP molecules.

Metabolic diseases result from the loss of any one of a number of enzymes required for efficient metabolism of various substances such as carbohydrates, amino acids and fats. These diseases involve either the accumulation of toxic substances in body, or else a deficiency in specific biological compounds. The majority of these diseases show strict recessive inheritance, i.e. they are caused by a single gene and one defective copy of that gene must be inherited from each parent.

Carbohydrates consist of molecules known as sugars (also known as saccharides, Gr. *sacchar*; sugar), some of which are simple while others are more complex. For example, sucrose, found in table sugar, is a disaccharide made up of just two monosaccharides sugars called glucose and fructose. Other carbohydrates, such as found in bread and pasta, consist of long chains of simple sugar molecules called polysaccharides. All these must be broken down by the body into their component monosaccharides by specific enzymes before they can be used to produce energy. A lack of any of these enzymes can lead to defects affecting carbohydrate metabolism which are generally inherited in an autosomal recessive pattern. For example, glycogen consists of a long chain of glucose monosaccharides which are the body's main source of energy. Any glucose molecules not immediately used for energy are held in reserve in the liver, muscles, and kidneys in the form of glycogen and released when needed by the body. Lack of a particular enzyme involved in linking glucose molecules to form glycogen or else breaking down the glycogen, can cause diseases known as glycogen storage diseases. These affect about 1 in 20,000 and while some of these diseases cause few symptoms, others are fatal. Low levels of sugar in the blood cause weakness, sweating, confusion, and sometimes seizures and coma. Treatment for some people can involve eating many small carbohydrate-rich meals to prevent blood sugar levels from dropping. Galactosemia is a disease caused when an individual is born lacking an enzyme necessary for metabolizing the sugar galactose which is a component of lactose (Lat. *lac*; milk), the main sugar found in milk. This leads to a toxic metabolite building up in the liver and kidneys which damages these organs and also damages the lens of the eye. An infant with galactosemia appears normal at birth but becomes anorexic and jaundiced and vomits heavily within a few days or weeks of consuming breast milk or lactose-containing formula. Without treatment this severely retards the physical and mental development of the infant and in the worst cases can be fatal.

Mammals have evolved a devious trick to shorten the length of time an offspring is dependent on the mother. This involves a molecular mechanism in which the gene for lactase, needed for digesting milk, is turned off at the end of the weaning period. Lactase is an enzyme needed for the metabolism of the lactose sugar found in milk. So while young mammals are able to digest milk from the mother they will reach a stage shortly after weaning when the consumption of milk will incur abdominal cramps and diarrhoea. This also occurs in humans; a child's bodily production of the lactase enzyme decreases dramatically after the first couple of years of life, resulting in lactose intolerance. Almost three quarters of the global population is lactose intolerant. This is one of the main reasons behind the production of mature cheeses in which a high proportion the lactose in the milk becomes degraded rendering it safe to eat. Most of the population of Europe and the US would still be milk intolerant if it were not for one or more individuals living around 4,000 BC in Sweden or the Middle East who acquired a particular mutation on chromosome 2. This has the effect of bypassing the normal shutdown in lactase production and so enabling the lifetime consumption of milk. This mutation proved to be a considerable evolutionary advantage following the first domestication of animals since milk then became a valuable source of nutriment. Most of Western Europeans have inherited this mutation and can safely consume milk, while many in African and Asian countries remain lactose intolerant. After returning from the *Beagle* in 1836, Charles Darwin suffered for over 40 years from long bouts of vomiting and intestinal pain, among other undiagnosed ailments. These stomach problems would begin two to three hours after a meal, the time it takes for lactose to reach the large intestine. Furthermore, his family history seems to show a pattern of inheritance of this condition suggesting a genetic origin. Darwin only got better when, by chance, he stopped taking milk and cream. European breeds of cats are also often lactose tolerant, in contrast to all other cat varieties, wild and tame. It may be that lactose tolerance was inadvertently selected for by Europeans who may not have understood that cats would normally not be able to digest milk and cream.

Amino acids, of which there are 20 different types, are the building blocks of proteins. Therefore, individuals with deficiencies in the enzymes needed for the degradation of amino acids are unable to break down certain components of protein. This can lead to the occurrence of severe brain damage caused by the toxic effects of the accumulating amino acids and their intermediates. These diseases are usually autosomal recessive inherited and will usually produce symptoms early in life. In most developed countries, newborns are routinely screened for several of the more common amino acid metabolism diseases, such as phenylketonuria and maple syrup urine disease.

Phenylketonuria results when individuals inherit mutations in the phenylalanine hydroxylase gene which produces the enzyme needed to convert the amino acid phenylalanine into the amino acid tyrosine. The resulting higher levels of phenylalanine in the blood harm brain development and can lead to mental retardation if not treated. In addition, some of the excess phenylalanine is converted into a phenylketone called phenylpyruvic acid, which, when excreted in the sweat and the urine, gives a musty odour, hence the name for this condition, i.e. phenyketones in the urine.

Two siblings born with phenylketonuria; the untreated bother on the left and his sister who received the restricted diet.

Perhaps one of the first reports of phenylketonuria was by the Nobel prize-winning American author, Pearl Sydenstricker Buck, in her book *"The Child Who Never Grew"*. Here she described how her 3 year old daughter that she brought from China to the US in the 1920s started to become severely mentally retarded in early childhood. She wrote: *"Although the child was beautiful, her mind was not developing. I remember when she was only 3 months old she lay in her basket on the sun deck of a ship. I had taken her there for the morning air. The people who promenaded on deck often stopped to look at her, and my pride grew as they spoke of her unusual beauty and of the intelligent look of her deep, blue eyes. I do not know at what moment the growth of her intelligence stopped."* A decade later, the Norwegian physician, Ivar Asbjørn Følling, also noticed the same disease in children living near his hometown of Oslo. Seemingly healthily growing infants would, at some point, stop developing normally and drift into irreversible mental retardation. He wondered whether the disorder might have something to do with metabolism and so set about seeing if there might be any unbroken down products in the urine of the children. He found that the urine of all affected children contained phenylketone which is a derivative of the amino acid phenylalanine. Many years later Pearl Sydenstricker Buck also recalled the characteristic musty odour now associated with the disease. Følling's discovery served two life-saving purposes: it allowed the diagnosis of this disorder, and also provided the cure, which simply involves restricting the diets of those affected children to only those foods containing low levels of phenylalanine. These days, thanks to Følling's pioneering work, dietary restrictions routinely form part of the treatment for amino acid metabolism disorders. Although the average incidence of PKU is about 1 in 15,000 births, this varies widely among different human populations from fewer than one in 100,000 births in the Finnish population to 1 in 4,500 births among the Irish and some other European populations where as many as 2–3% are heterozygous carriers of the mutant PKU gene. However, whether the distribution of the mutation in human populations reflects some unknown selective advantage or whether it is attributable to genetic drift remains a controversial issue. Genetic drift is used to describe random fluctuations in the frequency of the appearance of a gene in a population due to chance rather than natural selection.

Characteristic odours underlie a number of diseases resulting from inherited defects in the metabolism of amino acids. Another common disease is maple syrup urine disease. Characterised by urine with an odour resembling maple syrup due to the excretion of branched-chain amino acids, this disease results from defects in genes coding for enzymes involved in the metabolism of the amino acids leucine, isoleucine, and valine. The accumulation of these amino acids, along with their derivatives, leads to mental retardation in a similar way to phenylketonuria. Occurring in around 1 in 225,000 births, though as frequent as 1 in 176 in some inbred populations such as the Mennonites in Pennsylvania, early diagnosis and dietary intervention by restricting intake of the particular amino acids can allow the child normal intellectual development.

Defects in the metabolism of another amino acid, methionine, leads to a distinctive dried hops odour and is consequently often referred to as oasthouse (a building for drying hops) urine disease. Other kinds of metabolism disorders can also lead to a range of distinctive odours evident in urine, breath, or sweat. These include hypertyrosinemia (or rancid butter syndrome), cat' urine syndrome, isovaleric academia (with an odour of sweaty feet), and trimethylaminuria (fish odour syndrome). A very clear example of how mutations in specific genes can affect metabolism of particular chemicals in foodstuffs is provided by the vegetable asparagus. Asparagus contains a sulphur compound called mercaptan, also found in rotten eggs, onions, garlic, and skunk secretions, which when broken down by a specific enzyme in the digestive tract releases by-products causing a characteristic scent similar to rotten cabbage. The process is so quick that urine can develop the distinctive smell within 15 to 30 minutes of eating asparagus. Roughly half of the population in the UK have the gene for the enzyme which is supposedly inherited as an autosomal dominant trait.

Lipids (Gr. *lipos*; animal fat or vegetable oil), commonly known as fats, are an important source of energy for the body. Fat in the body is constantly broken down and reassembled, depending on the body's energy needs, by a range of enzymes. Defects in any of these enzymes can lead to the build-up of specific fatty substances that would normally have been broken down and which can be toxic. These are known as lipid storage disorders, of which one major example is Tay-Sachs disease. This disorder occurs due to mutations disrupting the function of an enzyme needed to break down specific lipids which then accumulate in tissues, particularly neurons that use this type of lipid for membranes and synapses. Children with this disease become progressively retarded, with associated spasticity, paralysis, dementia, and blindness, ending with an early death, usually by 4 years of age. Autosomal recessive inherited, Tay-Sachs disease is most common in families of Eastern European Jewish origin, where as many as 1 in 45 are carriers of the mutant gene compared to the general incidence of 1 in 300. This disease has actually been described in the Ashkenazi Jewish population as early as the 15 century. This high incidence suggests a possible heterozygous advantage in this population. There is some evidence that carriers of the Tay-Sachs disease gene have an increased resistance to Tuberculosis, and one idea is that Jewish populations in the past were often confined to urban areas high in *Mycobacterium tuberculosis* bacteria, and were consequently placed under selective pressures to evolve resistance to the disease. This high prevalence of Tay-Sachs disease, and other genetic diseases, in some Jewish communities has lead to the establishment of the Committee for Prevention of Genetic Diseases to develop routine genetic testing of young Jewish people worldwide. This was started in the 1980s by Rabbi Joseph Ekstein, who lost four children to Tay-Sachs disease. People tested are given a telephone number and a PIN, but are not told their results. Then, when a *shidduch* (introduction of marriageable singles) is suggested, the candidates can phone the organisation, enter both their PINs, and find out whether their union could result in severely disabled children. Although sometimes receiving criticism, such practices have led to a sharp decline in the occurrence of Tay-Sachs disease in Jewish communities.

Purines and pyrimidines are the bases found in the nucleotides making up DNA and RNA; Adenine and Guanine are purines whereas Thymine and Cytosine are pyrimidines. These are broken down in different pathways; pyrimidines into beta-amino acids and ammonia, while purines are degraded into urate which is further processed and excreted as uric acid. Inheriting mutations leading to a loss of an enzyme important in these pathways can lead to severe disorders. One such example is Lesch-Nyhan syndrome (LNS), where mutations in a gene (hypoxanthine-guanine phosphoribosyltransferase) on the X-chromosome involved in purine metabolism leads to increased levels of uric acid. This results in arthritis and severe neurological problems including mental retardation, and self-mutilating behaviour involving biting of the lips, tongue and fingers. LNS was first described in 1964 by Dr. Michael Lesch, a medical student, and Dr. William Nyhan, a paediatrician, at Johns Hopkins Hospital, Baltimore, when they identified LNS in two affected brothers aged four and eight.

Affecting about one in 380,000 births, LNS occurs mostly in males, as it is an X-linked disorder.

Another symptom of LNS is severe gout as high concentrations of urate in the blood can lead to uric acid crystals forming inside joints, often in the big toe, producing an intense pain; the word gout possibly derives from the Greek "*podagra*" (Gr. *pod;* foot, and *agra;* trap). Although most cases of gout appear to be the product of diet, mutations in certain genes involved in purine metabolism may lead to inherited predispositions to gout. For example, one such gene coding for an enzyme degrading uric acid, when mutated, can lead to an increased predisposition for gout and renal failure. This may explain why certain populations, such as people of the Pacific Islands, and the Maoris of New Zealand, show higher occurrence of this ailment. Incidences of gout in 3,000 year old skeletons on some Pacific islands point to the possibility of a founder effect; the genes responsible for this condition may have been carried to the islands by the early settlers. One possible case of familial gout was Charles V of Spain, one of the most powerful rulers of all time during his reign from 1516-1556. Charles V's gout first became evident when he was 28. As he grew older the attacks of gout increased in frequency and severity. Towards the end he had to be carried around in a specially designed chair as he was barely able to walk or use his hands. Recent analyses of his mummified body revealed high levels of urate deposits confirming his gout. His condition must have influenced many of his decisions ultimately affecting the destiny of many European countries. Charles passed on his vast empire, and his ailment, to his son Phillip. Although, also suffering severely from gout, Phillip supposedly rarely complained about his condition which he regarded as punishment from God for "*not being diligent enough to eradicate the protestant heresy*".

Charles V. Titian, 1548.

Lysosomes (Gr. *lysis*; loosening/dissolving) and peroxisomes are small spherical bodies found inside cells, where they degrade various molecules, bacteria or worn out organelles like mitochondria. They can be thought of as the cell's waste disposal units A genetic failure to manufacture an enzyme in lysosomes, needed for the breakdown of lipids, glycoproteins (sugar containing proteins) or mucopolysaccharides, can lead to the development of a lysosomal storage disease. In general, the cells that are most severely affected by this are neurons, because they contain large amounts of these substances, and once these cells are lost they cannot be replaced. The most common lysosomal storage diseases are a group of disease known as mucopolysaccharidoses caused by a failure to synthesize the enzymes needed to break down mucopolysaccharides, leading to various degrees of mental retardation and blindness in early childhood. Peroxisomes are another type of cell structure that break down toxic substances in the cells of the liver, kidneys, and brain. Diseases that disrupt the function of peroxisomes can lead to a high iron and copper build up in blood and tissue. The most severe of these is Zellweger syndrome which, with a frequency of 1 in 50,000 to 100,000 births, results in mental retardation and liver disease, with most affected infants dying before the age of 1 year old.

Chapter 13

RENAL, LIVER AND DIGESTIVE DISORDERS

It's what you do for your mates.

Grant Kereama, Jonah Lomu's kidney donor.

Kidneys form two bean-shaped organs, except in around 0.2% of the population where the kidneys are fused into a single horseshoe-shaped structure; a condition which the actor Mel Gibson has. The kidney's main job is to filter waste products and excess fluid from the blood, excreting this as urine.

Early childhood Kidney diseases often cause short stature due to the imbalance of salts and acids reducing appetite and preventing calcium being deposited in bones. Gary Coleman, most famous for the role of Arnold Jackson on the American sitcom "*Diff'rent Strokes*", was born with a congenital kidney disease causing nephritis (an autoimmune destruction of the kidney), which restricted his growth leading to his shortened stature of 142 cm.

The ailments of *Tiny Tim* in Dickens's novel "*The Christmas Carol*" is suggestive of renal tubular acidosis. This condition results in short stature, crippling weakness, kidney stones and untimely death if left untreated. Caused by the kidneys failing to excrete acids into the urine, the treatment generally includes sodium bicarbonate to neutralize the acids in the blood. Dickens had a keen interest in medical disorders and included many characters in his novels, with ailments that he had observed among friends and acquaintances. *Tiny Tim* was supposedly based on the invalid son of a friend who owned a cotton mill in Ardwick, Manchester.

In order to perform the task of filtering waste products from the blood, the kidney uses special membranes which act as selective barriers. Defects in these membranes can lead to kidney failure characterised by the presence of high concentrations of blood in the urine. This symptom occurred in the US president, Chester Alan Arthur, in 1882, along with fatigue and irritability. Shortly later he was diagnosed with kidney disease, a fact that was kept secret by Arthur's staff. Knowing that he would never survive a second term in office, Arthur did little to support his nomination in 1884 and died 2 years later. Defects in a gene on the X chromosome, coding for a protein important in this membrane formation can lead to such symptoms. Known as thin basement membrane disease, or Alport syndrome in more severe cases, this disease was first described by Dr. Cecil Alport in 1927 in a British family, in which a number of males died early of kidney problems while the female siblings remained unaffected.

Chester Alan Arthur, circa 1881.

One of the most common life-threatening genetic disorders is considered to be polycystic kidney disease (PKD). Affecting around 12.5 million people worldwide, (1 in 500 to 1 in 1,000 individuals) large numbers of cysts develop in the kidney which can eventually shut down kidney function. Generally inherited in an autosomal dominant pattern, two genes are so far known, (PKD1 and PKD2), that are involved in cell signalling. One idea is that when either of these is mutated the cells in the kidney may grow and divide abnormally. Although, it is possible that Hippocrates might have described the disease, the first documented case of PKD appears to have occurred in Stefan Bathory, King of Poland who died in 1585 at the age of 53. The surgeon who performed an autopsy described the kidneys as *"large like those of a bull with an uneven and bumpy surface"*. The death of the American Humorist and news column writer, Erma Louise Bombeck, in 1996, raised the profile of this inherited condition in the US.

The Russian writer Mikhail Bulgakov died of kidney disease in 1940, suffering from the same symptoms as his father and dying at the same age of 48. The symptoms suggest that he may have inherited from his father a dominant gene for nephrosclerosis, a condition in which high blood pressure leads to a thickening of arteries in the kidney, reducing its ability to function properly. Educated as a medical doctor, Bulgakov almost certainly knew the implications of his failing health. He therefore focused the last months of his life on trying to finish his final novel. This was his political and satirical masterpiece, *"The Master and Margarita"*, which was critical of the Stalin regime. He had destroyed several previous drafts of the novel for fear of arrest by the soviet secret police. It seems very likely that realisation of his impending death prompted him to finally publish his novel.

Mikhail Bulgakov, 1939.

Is life worth living? It all depends on the liver.

William James

The liver is found in the upper right-hand side of the abdomen under the ribs and weighs about 1.5 kg. The liver receives blood directly from the intestines containing almost everything absorbed by the intestines, including nutrients, drugs, and sometimes toxins. This flows through a latticework of tiny channels inside the liver, where the digested nutrients and toxins are processed. Blood also enters from the heart carrying oxygen to the liver tissues as well as cholesterol and other substances for processing. Many different jobs are carried out by the liver cells (known as hepatocytes, *hepar*; liver): they produce and excrete bile required for food digestion, perform several roles in carbohydrate metabolism and lipid metabolism, produce coagulation factors, and break down toxic substances and haemoglobin. The breakdown of haemoglobin results in the production of bilirubin which is further converted into bile - an increased build-up of bilirubin causes the orange colour typical of jaundice and many liver diseases. The liver is the only organ in the body which can regrow if part of it is removed. This allows surgeons to be able to cut out cancerous liver leaving a small amount behind to regenerate. In Greek mythology, *Prometheus* was punished by the gods for revealing fire to humans by being chained to a rock where a vulture would peck out his liver, which would grow again overnight only to be eaten again when the eagle returned the following morning. How the ancient Greeks knew about the liver's regenerative capacity is a mystery.

When chronic diseases cause the liver to become permanently injured and scarred, the condition is called cirrhosis. This is the fifth highest cause of death in the UK responsible for between 5,000 and 10,000 deaths per year - the majority due to excessive alcohol consumption. George Best died from this in 2005 as a result of his alcoholism with the British tabloid *News of the World* carrying out his final request of publishing a picture of him in his hospital bed along with a message: *"Don't die like me"*.

Prometheus. Peter Paul Rubens, 1612.

Less common causes of cirrhosis include direct liver injury from genetically inherited diseases such as cystic fibrosis, alpha-1-antitrypsin deficiency, galactosemia, and glycogen storage disease, as previously described. Two further inherited disorders result in the accumulation of metals in the liver leading to tissue damage and cirrhosis. In Wilson's disease too much copper is stored, and in haemochromatosis too much iron is absorbed depositing in the liver and other organs.

Wilson's disease is an inherited autosomal recessive disorder characterized by the pathological accumulation of copper in the body. Affecting around 1 in 30,000 individuals this is caused by mutations in the Wilson disease protein (ATP7B) gene. Around 1 in 100 people carry a single abnormal copy of this gene. The liver, unable to deal with the copper, releases it into the blood where it deposits throughout the body, particularly in the kidneys, eyes and brain leading to psychiatric symptoms in addition to liver disease. Augustus Caesar is known to have suffered chronic ill health with symptoms of intestinal and immune system disorders suggestive of a syndrome affecting the liver and kidneys, possibly Wilson's disease. Though he still lived to the age of 75, there is some suggestion that he might have passed the gene for this disease on to two of his descendents, Nero and Caligula, who also showed mentally instability and epilepsy. The Claudio-Julian dynasty of the early Roman Empire carried a number of inherited familial ailments, which was almost certainly a result of intermarriage. Epilepsy, for example figured a great deal in the Julio-Claudian dynasty with the great Julius Caesar suffering a number of well documented seizures during his reign. It is possible that the sudden deaths of his father and great grandfather might have been as a result of epileptic episodes.

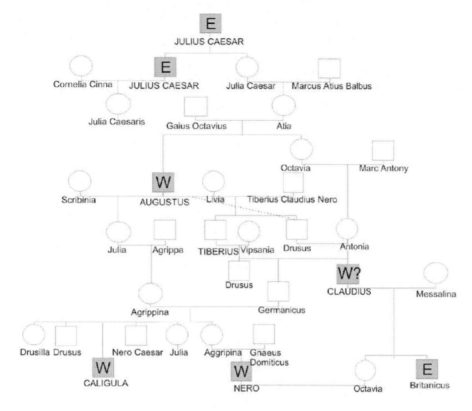

The Julio-Claudian family tree. Grey boxes signal the possible occurrence of either epilepsy (E) or Wilson's disease (W).

Ernest Hemingway, 1939.

An average person has around one gram of iron in their body, whereas a sufferer of haemochromatosis can store around twenty grams. There are some suggestions that this might even be enough to set off airport metal detectors. The increased iron accumulation in this inherited liver disease, particularly the liver pancreas and heart, can lead to liver damage, diabetes mellitus and congestive heart failure. Haemochromatosis is one of the most common genetic diseases, affecting as many as 1 in 200 among Europeans, with as many as 1 in 10 people carrying a mutation in one of the genes regulating iron metabolism. This high incidence has led to the suggestion that the disorder may have, or previously have had, some selective advantage deriving from the higher iron levels in the body.

Most patients with haemochromatosis show no symptoms until they are over 50 years old, at which time they may present with symptoms of arthritis, diabetes, fatigue, an enlarged liver, and other symptoms such as pigmentation of the skin due iron deposition and melanin in the skin; for this reason it is often referred to as bronze diabetes. Clarence Hemingway, the physician father of Ernest, is known to have suffered from this condition prior to his suicide at age fifty-nine. Interestingly, the excess iron is also suspected of leading to depression or instability in the cerebrum area of the brain and may explain some of his family's history with three of his six children, including Earnest, together with a granddaughter, all suffering from diabetes and depression leading to suicide. Another possibility is that Earnest and his siblings might have also inherited bipolar disorder for which Ernest received electro-convulsion therapy. Blaming this therapy for robbing him of his memory, he took his own life just short of his 62nd birthday, in 1961, with a shotgun blast to the head. Judged not mentally responsible for his final act, he was allowed a Roman Catholic funeral. This syndrome has also been reported in the family of another famous American author, John Steinbeck, whose son John Steinbeck IV was affected. "*I inherited two life-threatening diseases from my parents*", he explained, "*haemochromatosis and alcoholism*". After his death in 1991, his wife published a biography which detailed how he managed to cope with the two illnesses.

Some people are cold by temperament; that may be a misfortune for them, but it is no more a sin than having a bad digestion is a sin.

CS Lewis

It appears that there are a number of digestive maladies that affect the Jewish population in disproportionate numbers: Crohn's disease, irritable bowel syndrome, some food allergies and even lactose intolerance have all been associated with Jewish populations. Dr. Burrill Crohn, after whom the disease is named, recognized Crohn's as a particularly Jewish disease, reputedly joking to other doctors that the ailment was brought on either by Jewish genes, Jewish food or Jewish mothers. Resulting in abdominal pain, diarrhoea, vomiting and weight loss the disease is caused by an overreaction to bacterial flora due to a lack of specific antibiotics, known as defensins, which are secreted by the intestines. One gene known to cause this condition codes for one of these defensins. Dwight Eisenhower, the 34th President of the United States was diagnosed with Crohn's disease in 1956, six months before the presidential election. Once again, presidential staff chose not to disclose this information to the public until a severe inflammation of his intestine forced him to undergo surgery. Recovering from this, he won a second term as president.

Dwight David Eisenhower, 1956.

Some digestive disorders are characterised by overreactions to specific products such as lactose, which was mentioned earlier. Coeliac disease, results from exposure to gluten and related proteins, which are found in wheat, rye, malt, barley and oats, and leads to damage of the lining of the small intestine and a reduced ability to absorb nutrients. Though not strictly inherited, there appears to be a genetic component in many childhood cases with the nutritional deficiencies causing a range of symptoms such as diarrhoea, weight loss stunted growth and fatigue. Joe C, the sidekick of rock star Kid Rock, suffered from coeliac disease, sadly dying from its resulting complications at the age of 26.

Intestinal disorders presenting in newborns can be extremely dangerous if not recognised immediately, particularly if a newborn fails to pass a stool properly and is reluctant to eat. Acting couple Denise Welch and Tim Healy's son Louis was born with a disease known as Hirschsprung's disease. Affecting one in 5,000 babies, the disease is caused when nerve endings in the bowel do not form properly hindering the passage of digested food and stool. This may require an operation involving removing the affected part of the intestine lacking the normal nerve cells; Louis had a part of his large intestine removed at the age of 6 weeks and has made a full recovery. Hirschsprung's disease affects males much more frequently than females and there is a 3 to 12 percent chance that a couple with one affected child will produce another baby with the disease. Both these trends suggest a genetic component to the condition and indeed, a number of mutated genes that play roles in the development of neural cells in the digestive tract during embryonic development, are known to contribute to Hirschsprung's disease.

Chapter 14

VISUAL DISORDERS

Painting is a blind man's profession. He never paints what he sees, but what he feels.

Pablo Picasso

Eye colour, is determined by the levels of melanin pigment in the iris (Gr. *iris*; rainbow). The iris is the circular coloured curtain of the eye which, opening to form the pupil, regulates the amount of light entering the eye. The melanin is produced here to absorb excess light too strong or bright for vision, with the differing amounts determining the colour of a person's eyes; blue resulting from low levels and brown from high. Caucasian babies are often born with blue eyes because at the time of birth they have not begun to produce melanin in their irises, in contrast to babies of other ethnic origins such as in Africa and Asia who are generally born with dark eyes. Previous assumptions were that the inheritance of eye colour was relatively simple involving only one gene with a version (allele) for brown eyes dominant over blue eyes. However, the genetic basis for eye colour is actually far more complex and is a polygenetic trait i.e. governed by the presence of more than one gene. At least 6 genes are known to interact with each other to give the various colour shades and so almost any parent-child combination of eye colours can occur. The most common eye colour is brown, while blue eyes are found mainly in people of Northern European and Eastern European descent. Green is one of the rarest eye colours, though is fairly common among Pashtuns (e.g. Sharbat Gula the "Afghan Girl" from the June 1985 cover of *National Geographic*) as well as people of Celtic, Germanic and Slavic descent.

One relatively common phenomenon is the presence of two differently coloured eyes in an individual. Known as heterochromia, as many as 1 in a 100 people are affected where either both eyes are separate colours (complete heterochromia) or there are a variety of colours within a single iris (sectoral heterochromia). A number of movie actors have eye colour irregularities, the presence of which may play some roles in creating a more striking appearance. For example, Christopher Walken has one blue eye and one hazel eye and Jane Seymour has one green eye and one brown eye. Two people with sectoral heterochromia are Kate Bosworth who has a hazel section at the bottom of her right, otherwise blue, eye and Jessica Cauffiel who has brown pigmentation in her left eye. Heterochromia is also particularly common in dogs such as Dalmatians and Border Collies. David Bowie supposedly does not have heterochromia - his discoloured eye is actually a permanently dilated pupil that he sustained after getting punched in a fight over a girl when he was 13.

Jane Seymour.

Heterochromia is frequently genetic in origin, and often autosomal dominant inherited. In addition there are some inherited syndromes which present with this feature in a combination with other symptoms, for example Waardenburg syndrome, which is caused by mutations in a developmental gene (pax3) leading to pigment disturbances in the iris, hair and skin, as well as hearing loss. Piebaldism and Hirschsprung's disease also sometimes present with heterochromia.

Dr. William Archebald Spooner (1844-1930).

Albinism, resulting from defects in melanin production also leads to visual defects. Affected individuals generally show sensitivity to light due to the lack of any melanin in the iris absorbing any stray light entering the eye. However, the major vision defects in people with albinism result from an abnormality in an area of the retina important for sharp vision such as reading. This tissue requires melanin to properly develop nerve connections between the retina and the brain. When these connections are disrupted involuntary movements of the eyes, known as nystagmus, can occur in addition to strabismus where the eyes are unable to fixate together, i.e. 'cross-eyed'. Dr. Spooner, the famous Oxford classicist, may have had this condition. It is possible that nystagmus, associated with his albinism, caused a jumbling of information from the printed page which led to his famous speech aberration. Strabismus, affecting around 5 percent of people, can lead to a loss of binocular vision and depth perception. A number of the world's most famous artists may have actually owed their talents, to some degree, to this condition. Rembrandt's divergent squint, which is clearly evident in his self portraits, has led to the suggestion that this contributed to his ability to translate a 3-dimentional world to a 2-dimentional canvas. Albrecht Durer, who has been credited by some have credited with formalising and refining linear perspective in Renaissance Art, is another artist who suffered congenital strabismus, as did the great French artist Edgar Degas who once commented to a friend: *"One sees what one wants to see and the falsehood constitutes art"*.

Self Portrait at 26. Albrecht Durer, 1498.

Myopia (short-sighted vision) is the most common eye problem in the world. Affecting a quarter of the population, it occurs when the lens of the eye is unable to flatten properly resulting in light entering the eye to focus in front of the retina. Myopia is hereditary to a large degree, and a number of genes are thought to be responsible, though a mutation in one gene (pax6) is particularly influential in eye development. Severe mutations can lead to diseases characterised by a massive underdevelopment of the eye while less disruptive alterations in this same gene can result in myopia with the eye adopting a slightly elongated front to back shape which affects focusing. Myopia usually presents during the pre-teen years, worsens gradually as the eye grows during adolescence, and then levels off as a person reaches adulthood. Another visual anomaly occurs when the front part of the eye is not perfectly spherical. Known as astigmatism, the eye has different focal points in different planes, i.e. an image may be clearly focused on the retina in the horizontal plane, but in front of the retina in the vertical plane. As with myopia, this condition can be corrected with lenses. It has been suggested that the famous elongated figures in the paintings of the Cretan artist El Greco were due to his astigmatism. Judging from their painting styles, astigmatism may also have afflicted a number of other artists including Holbein, Botticelli, Titian, and Modigliani.

Saint Martin and the Begger. El Greco, 1599.

Visual defects appear to be strikingly common among artists in general. In a study of the teachers and students of the École des Beaux Arts in 1917, the incidence of myopia (i.e. short-sighted vision) was found to be 48 percent, a figure nearly twice that of the general population. Individuals with myopia see objects near to them, while more distant objects appear blurred. Therefore, an artist with myopia might put more emphasis on the shapes of objects at the expense of detail, and so it is therefore tempting to speculate that the development of Impressionism - an art style with emphasis on contour and colour at the expense of detail – developed as a result of myopic vision. Artists of this Impressionist period who are known to have been myopic include Renoir, Cezanne, Pissarro, Degas, Dufy, Derain, Braque, Vlaminck, Rodin, and Matisse. Interestingly, Renoir and Cezanne both refused to wear glasses - Renoir often stepped back from his canvas to further enhance the impressionist effect.

Some visual disorders result from neurological conditions in which the brain's interpretation of visual perception is affected. One such condition leads to temporary attacks of distortion. It is called Alice in Wonderland Syndrome after the title character in the novel by Lewis Carroll who finds herself growing and shrinking during her adventures. During an attack a door knob, for example, can appear to be the size of the door itself or the flat floor can appear to slope upwards and then drop off entirely. The majority of patients with this syndrome have a family history of migraines, and perhaps not coincidentally, Lewis Carroll suffered himself from severe migraine.

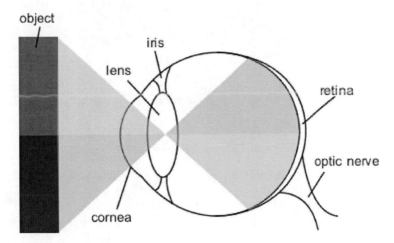

A diagram of the eye.

It is the eye of ignorance that assigns a fixed and unchangeable colour to every object; beware of this stumbling block.

Paul Gauguin

Light entering the eye is focused by the lens onto a light-sensitive panel of cells known as the retina at the rear of the eye. It is here that the light is detected and converted into electrical signals by photoreceptor cells which then transmit to the brain via the optic nerve. There are two types of photoreceptor cells in the retina referred to as rod cells and cone cells: cone cells are responsible for sharp vision and colour vision while rod cells (named for their cylindrical shape) are highly sensitive to light and so are responsible for night and peripheral vision. There are many different diseases of the retina, affecting the photoreceptor cells. While some gene defects result in the absence of both types of photoreceptor cells, others preferentially affect either the rod or the cone cells. Leber's congenital amaurosis, for example, is an inherited defect in both rod and cone cells and so characterised by blindness at birth. This disease affected a now very famous dog by the name of *Lancelot*. Totally blind at birth due to the inheritance of a gene for this disease, he was injected with a genetically engineered virus into his eye which carried a healthy copy of the gene to his retina. As a result, *Lancelot* can now catch a ball and has given hope for the possibility of sight restoration through gene-therapy.

An uncontrolled growth of the retina, leading to a cancer of this tissue, is caused by the inheritance of mutations in a gene called the retinoblastoma-1 gene. This gene produces a protein that functions in normal cells as a tumour suppressor acting as a brake on cell division. This disease, known as retinoblastoma affects either one or both eyes in around 1 in 15,000 children. Peter Falk, the American actor, best known for his role as *Lieutenant Columbo*, suffered from retinoblastoma as a child losing his right eye at the age of three and consequently wears a glass eye.

Lancelot, who was blind since birth, had his sight partially restored through gene therapy.

Defects affecting only rod cells of the retina cause night-blindness where individuals find it difficult to see in the dark. This is a symptom of several heritable eye diseases, the most common being retinitis pigmentosa (RP), which affects around 1 in 3,000 people. Retinitis pigmentosa (RP) is one of the most common forms of inherited retinal degeneration typically characterized by the progressive loss of rod cells leading to night-blindness followed later by a reduction of the peripheral visual field (known as tunnel vision). Presently, 35 different genes are known to cause RP when mutated and consequently the disease shows many different patterns of inheritance depending upon which gene is affected. Former child actor Isaac Lidsky who played the role of *Weasel* on *"Saved by the Bell: The New Class"* was diagnosed with the disease at age 13. Two of his sisters also inherited the same gene and also have the same condition. Giving up acting, he studied Applied Mathematics and Computer Science at Harvard College and in 2008 became the first legally blind US Supreme Court clerk. One of Mexico's biggest music stars, Rigo Tovar, suffered from RP. Losing his sight in his 20's, he would constantly wear dark sunglasses to protect his retina, although many people did not understand he suffered from RP believing it to be an affectation. Another kind of night blindness is known as nyctalopia (Gr. *nyct*; night, *aloas*; obscure/blind, *opsis*; vision) and can result from vitamin A deficiency - this is metabolized to retinol which is important for the functioning of the retina. The ancient Egyptians in the sixteenth century BC first described the cure for this which involves ingestion of ox liver, a food rich in vitamin A. Not until 1913 was the role of vitamin A deficiency in nyctalopia fully recognised and, sadly, over 3,500 years after the ancient Egyptians were treating the disease, it is still estimated that 70 percent of the 500,000 pre-school children in the world who develop blindness do so as a result of vitamin A deficiency.

Cone cells are less sensitive to light than the rod cells, but allow the perception of colour and fine details. Defects in cone cells result in colour blindness and hemeralopia, often known as day blindness where an extreme sensitivity to light can occur. One inherited disease characterised by this is Stargardt's disease. Here mutations in a gene, (ABCA4), producing a protein involved in providing energy to retinal cells, leads to the gradual deterioration of cone cells and central vision, along with the ability to perceive colours, while leaving peripheral vision intact. Inherited as an autosomal recessive trait, the symptoms begin in childhood - affecting approximately one in 10,000 children – where there is wavy vision, blind spots, blurriness, impaired colour vision, and difficulty adapting to dim lighting. The athlete Marla

Runyan, though having competed in the Olympic final in Sydney 2000 for the 1,500 meters **run, suffers from Stargardt's disease. Her vision was perfectly normal** until she was about 9 years old at which time she began to experience problems seeing the blackboard in school and reading books. As a result of this condition Marla Runyan is legally blind but is still able to **see shadows and peripheral details. In her biography,** "*No Finish Line*", **she .recalls being** amazed to discover that other hurdlers could see all ten hurdles on the track, when she could barely see the first one! It was this that led her to switch to distance running and she still races competitively. In the 2010 Winter Olympics, the Canadian cross-country skier Brian McKeever who also suffers from Stargardt disease became the first winter athlete to compete in both the Paralympic and Olympic games.

Peter Falk.

Although some degree of colour blindness is relatively common, total colour blindness, known as achromatopsia, is a rare inherited condition. Many of the inhabitants of the small coral island of Pingelap, in the Western Pacific Ocean, are unable to distinguish any colours. A freak storm in the eighteenth century killed most of the islanders leaving only around twenty women and a handful of men including a young chief who happened to be a carrier of the gene for achromatopsia. Consequently, around a third of the atoll's 3,000 inhabitants now carry the gene leading to 6 percent of the population suffering from the disorder. Worldwide, by contrast, achromatopsia only affects around 1 in 33,000.

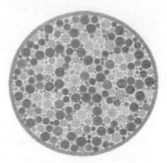

An Ishihara test for **red-green colour deficiencies. The number "10" should be visible** to people with normal colour vision.

Most people who are colour blind do not live their lives entirely in black and white as there are many different types and degrees of colour blindness which should be more correctly labelled as "colour deficiencies". This is because humans have three types of cone cells each containing a different photopigment which respond to variation in colour in different ways so the perception of colour results from a mixture of the outputs of all three. An absence of only one of the particular photopigments will lead to different variations of colour blindness. The most common deficiency is red-green colour blindness. This type affects around 8 percent of males, as the genes for two of the photoreceptors involved in this disorder are on the X chromosome. This results in a lack of perception of red, orange, green, blue and cyan. Many people are unaware of their problem which can often only be revealed by tests such as an Ishihara diagram. Alternatively, problems could be encountered in trying to accomplish certain tasks involving distinguishing colour, such as chemical titration. John Dalton, one of England's most renowned chemists, was red-green colour blind and was actually one of the first to describe the condition. He correctly surmised that it must be hereditary, since his brother had the same visual defect. Mistakenly though, he was convinced that this defect resulted from the lens of his eye being tinted blue, and he requested that after his death (in 1844) his eyes be removed and examined to confirm this. Even though after his death his lenses were found not to be tinted, for many years colour blindness was called Daltonism, after him. However, the potential problems associated with red-green colour blindness were not widely appreciated until 1875, when a train accident in Sweden was attributed to the driver's colour blindness.

Erroll Boyd, 1930.

Interestingly, while as many as 8 percent of North Europeans are affected by this disorder, there are considerably lower frequencies in other ethnic groups. One possible explanation is that individuals with colour blindness may develop better night vision. It is suggested, therefore, that the prevalence of colour deficiencies may associate with the duration of twilight; longer twilight periods, at higher latitudes, may have resulted in improved hunting skills for colour blind individuals in these regions. The famous Canadian World War I combat pilot and aviation daredevil, Erroll Boyd, was turned down by the Royal

Flying Corps because of his colour blindness. However, he was accepted by the Royal Naval Air Service, where he excelled as a night pilot. He also became the first person to fly across the North Atlantic, in 1930, outside the summer season. Later on, Boyd recalled: "*Boy, it was dark! I felt as though I was piloting a car in a coal mine.*"

The greyish-green paintings of John Constable have led many to suggest that the great landscape artist might have harboured a colour deficiency, along with other notable artists such as James Whistler, Charles Meryon, Fernand Leger and Piet Mondrian, all of whom also characteristically used more restricted pallets. Some artists, to compensate for their colour deficiency, have resorted to using the brightest colours possible. Royce Deans, whose paintings adorn the walls of a number of *McDonalds* restaurants, has cited this as one of the reasons behind his distinctive style.

Sailing boat. Lloyd Frederic Rees, 1982.

While the majority of colour deficiencies are of the red-green type, the loss of a different photopigment leads to a blue-yellow colour confusion. Occurring in 1 in 1,000 individuals it affects males and females equally. A similar visual defect can occur as a result of jaundice and cataracts. Lloyd Frederic Rees' famous paintings of Sydney harbour and Ayres Rock, in striking bright yellows and faint blues, have been attributed to the artist's cataracts.

It is the optic nerve that transmits visual information from the retina to the brain. Having no capability for regeneration, there can be no recovery from diseases causing degeneration of this nerve which can result in irreversible blindness in young adulthood. There are two types of inherited optic atrophy both of which effect the functioning of mitochondria, and so possibly disrupting energy utilization in the nerve tissues, in particular the optic nerve. One disease, known as Kjer's optic neuropathy, is inherited dominantly through any of several genes on various chromosomes which control the stability of mitochondria. The other, Leber's hereditary optic neuropathy (LHON), is a mitochondrial inherited disease due to a mutation of the mitochondrial genome and hence is passed exclusively through the mothers. Affecting only around 1 in 30,000 people in Europe, there is a disproportionate amount of French-Canadians sufferers in Quebec, where it is often referred to as the *Frenchman Disease*. This serves as another example of a founder effect. When Canada became a royal province in 1663, there was an unfavourable ratio of six male colonists to every European-born female. With a view to reducing this imbalance and to ensuring the stability of the colony, the French King Louis XIV devised a program whereby over 700 young single French women were sent out between 1663 and 1673. Most of these women were orphans who had already been placed under the care of the king and so became known as the "*filles du roy*," or "*King's daughters*". Many of the 5 million French-Canadians living in Quebec province are descendants of these women. One, however, carried a single nucleotide change in her mitochondrial DNA. Marrying one of the colonists in Quebec City in 1669, she produced five daughters who also all married in or near Quebec City. Today, as many as 90 percent of French-Canadians affected with LHON can trace their ancestry back to this woman.

Johann Sebastian Bach. Elias Gottlob Haussman,1746.

Another group of diseases of the optic nerve is Glaucoma. The definition of glaucoma has actually changed since its introduction around the time of Hippocrates in 400 BC where the word glaucoma derived from the Greek for blue, probably originally describing cataracts. Glaucoma is a term used for a group of diseases that can lead to damage to the eye's optic nerve and result in blindness. This generally occurs when the fluid pressure inside the eye is too high. Affecting around 1 in 200 people it does not generally exhibit Mendelian inheritance though relatives of people who have the condition have a significantly higher risk of developing the condition themselves. There are, however, a small number of cases of familial glaucoma where mutations in a number of key genes lead to a very high susceptibility, usually at an earlier age. The most common mutation is found in the GLCA1 gene on chromosome 1 which produces a protein involved in pressure regulation of the eye, the disruption of which can lead to increased intraocular pressure. It is this increased intraocular pressure that is the biggest risk factor for glaucoma and can lead to optic nerve degeneration. The disease can manifest itself as acute, where sight is lost very quickly, or chronic. The latter, often known as "*the silent sight thief*", leads to a gradual, often unnoticed, impairment of vision taking place over a long period of time. John Milton, considered one of the greatest poets of the English language, wrote his epic poem *Paradise Lost* (1667) after he became blind from glaucoma. It was while he was in his 30s, that he felt his sight getting weak and dull, which he falsely attributed to eyestrain from constant reading. By the age of 43 he was totally blind. However, he continued to write, and produce volume after volume of prose and poetry; 250 years later the great activist Helen Keller, who was both deaf and blind, set up the *John Milton Society for the Blind* to provide literature to deaf and blind persons.

It is has been argued that Milton's blindness stimulated his greatest work. "*He sacrificed his sight, and then he remembered his first desire, that of being a poet,*" stated the great Argentinean writer Jorge Luis Borges in one of his lectures. Jorge Luis Borges also lost his sight to glaucoma, which he had inherited from his father who had become blind in middle age, as had other relations on his father's side of the family. "*In my case, that slow nightfall, that slow loss of sight, began when I began to see. It has continued since 1899 without dramatic moments, a slow nightfall that has lasted more than three-quarters of a century. In 1955 the pathetic moment came when I knew I had lost my sight, my readers and writer's sight*". A century after Milton gradually lost his sight, composer Johann Sebastian Bach went blind in a violent flash. His symptoms suggest that he may have suffered from another form of glaucoma known as acute, or closed-angle, glaucoma, although he had also suffered from deteriorating vision for some years prior to this, which he blamed on a lifetime of copying music in the dim light of church organ lofts. Desperate to recover his sight again, he visited a surgeon and died months after a futile, and possibly harmful, operation on his eyes in 1750.

Chapter 15

HEARING DISORDERS

What matters deafness of the ear, when the mind hears. The one true deafness, the incurable deafness, is that of the mind.

Victor Hugo

Twenty years after the death of Bach, arguably the greatest composer of all time was born, Ludwig Van Beethoven. He would suffer from a different impairment in his later life - deafness. He produced some of his greatest works, such as *Ode to Joy* in 1824, after he became totally deaf; some believe his deafness enhanced his creative genius. It was at around the age of thirty that he began to suffer from buzzing in both ears. Sometimes this would clear for a few months at a time but it eventually ended in complete deafness. His autopsy on March 27th 1827, by a Dr. Wagner, discovered that his ear cartilage was large and misshapen suggesting that his hearing loss occurred through abnormal growth of bone in the inner ear, which is a characteristic of otosclerosis. This is the most important cause of chronic progressive hearing loss in adults and can be thought of as a mild form of osteogenesis imperfecta with mutations in collagen genes being shown to play roles in a number of cases. A number of other musicians have also suffered from this including Dina Carroll, and also Frankie Valli of the *4 seasons*. These days the condition can be corrected by surgery. In addition to the bones in his inner ears, a number of Beethoven's other bones also continued to overgrow. His head became increasingly large, with a prominent forehead, enlarged jaw, and protruding chin. It is not certain which of his many ailments (which also included lupus, typhus, syphilis and liver cirrhosis) eventually killed him. Shortly before his death he stated *"When I am dead, request on my behalf Professor Schmidt, if he is still alive, to describe my disease, and attach this written document to his record, so that after my death at any rate the world and I may be reconciled"*. The composers' name lives on in the *Beethoven Mouse* gene, which when disrupted results in progressive loss of hearing in a mouse; studies are under way to link this to humans.

Ludwig van Beethoven when composing the Missa Solemnis. Joseph Karl Stieler, 1820.

There have been a number of other composers who have written some of their greatest works after the onset of hearing difficulties, such as the Czech composer Bedřich Smetana and his 6-part masterpiece, *Má vlast,* and also the renowned Briton William Boyce who compiled and wrote great volumes of cathedral music during the 1700s. What is often not realised is that deaf people can also experience music by feeling the vibrations in their body; even people who have been born deaf have the ability to appreciate and make music. Perhaps the most renowned deaf musician in the UK is the Scottish award-winning percussionist, Evelyn Glennie, who lost her hearing at the age of 12. She set about learning to distinguish how different pitches and notes felt by placing her hands against the classroom wall while her percussion teacher played drums.

Around 1 in 1,000 newborns present with profound hearing impairment with genetic factors possibly accounting for as many as half. At present 120 genes have been found, that when mutated, are responsible for deafness. However, more than half of all recessively inherited deafness results from mutations in only one gene, known as connexion 26 (Cx26). Around 4 percent of the population carry a mutant version of this gene which normally produces a protein important in forming channels between cells to allow for the passage of small molecules and ions important in the inner ear. More severe disruptions of this same gene also lead to skin abnormalities. Individuals having only one copy of such a mutant gene tend to develop increasing skin thickness suggesting that a selective advantage could exist for this gene providing protection against possible cuts and infections, perhaps explaining why so many of the population still carry this gene variant.

Another possible reason for the high frequency of this gene mutation, and other deafness-causing genes, are the establishments of institutions for the deaf leading to intermarriage among the deaf - in the US a staggering 85 percent of individuals with profound deafness marry another deaf person. This is supported by the fact that 200 years ago Cx26 deafness was half as common. Alexander Graham Bell, whose mother and wife were both deaf, speculated that continued intermarriage among the deaf might someday result in the formation of a hearing-impaired race. This prompted him to suggest closing residential schools for the deaf in favour of mainstream education.

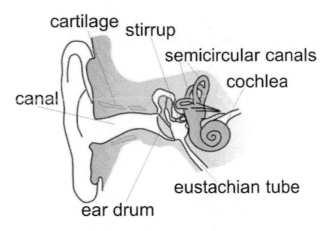

The structure of the ear.

In addition to single gene defects affecting only deafness there also nearly 400 syndromes in which deafness is a feature. Waardenburg Syndrome is one of the most common. Waardenburg syndrome is an autosomal dominant group of genetic conditions which account for around 1 percent of all deaf individuals. The disease is characterized by hearing loss and changes in pigmentation of the hair, skin, and eyes. People with Waardenburg syndrome often have very pale blue eyes, heterochromia, and distinctive hair colouring such as a patch of white hair often known as a witch's streak. The syndrome derives its name from the Dutch doctor Petrus Johannes Waardenburg who first noticed that people with differently coloured eyes often had a hearing impairment. Charles Darwin also noticed this same effect in cats observing that "*cats which are entirely white and have blue eyes are generally deaf*". Darwin was intrigued by how apparently unrelated characteristics such as eye and skin colour could be related to deafness. We now know that melanin, the dark pigment found in the skin and the eyes, also has a function in the development in the ear; Waardeburg syndrome results from mutations in certain genes effecting the production of this pigment. Another syndrome in which deafness is a feature is known as Jervell Lang-Nielsen syndrome. This results from defects in a gene producing a protein involved in potassium ion channels, which in addition to the ear, are also important for heart muscles. This disorder is often the main suspect in the rare cases of sudden death among deaf individuals. This was first noted in 1856 when a girl by the name of Steinin, at the Leipzig School for the Deaf, collapsed and died while being publicly admonished by the Director for a misdemeanour. When the parents were informed it emerged that two of the girl's deaf brothers had also previously died following episodes of emotional stress.

Around 24,000 people in the UK are both deaf and blind with the majority losing their senses later in life from infections such as rubella. The major genetic cause for deaf blindness is Usher syndrome which is recessively inherited from mutations in a number of different genes playing roles in the development of the inner ear and retina. With an incidence of 1 in 25,000, it is characterized by deafness at birth and a gradual vision loss during the first decade of life. One well known case of Usher syndrome was John Tracy, the son of the American actor Spencer Tracy and his wife Louise. It was through their son that they founded the *John*

Tracy Clinic in 1942, a private, non-profit education centre for infants and preschool children with hearing loss in Los Angeles, California, USA. It still provides free services worldwide for parents of children with hearing loss.

The Spanish painter Fransisco Goya suffered from an autoimmune disorder known as Vogt-Koyanagi-Harada syndrome which left him permanently deaf in 1792. In addition to the hearing loss he also suffered problems with balance and coordination. This is because the sensory organs for the detection of orientation and movement are also found in the inner ear. Isolated from others by his deafness, he became increasingly occupied with the fantasies and inventions of his imagination and it has been suggested that the intensely haunting themes of Goya's "*Black Paintings*", created in the later years of his life, were influenced by his disease.

Life is like riding a bicycle. To keep your balance you must keep moving.

Albert Einstein

The balance organs in the inner ear basically consist of tubes and cavities containing fluid and small hairs which are moved in response to motion of the head. Spinning around, can cause the fluid to keep sloshing around long after motion has ceased, with the hair cells still informing the brain that the body is still moving, even after it has stopped – this causes the dizziness. A similar situation can occur whilst reading in a moving car - the inner ear senses the movement of the vehicle, but the eyes see only the book which is not moving. The resulting sensory conflict can lead to the typical symptoms of motion sickness, such as nausea. *Nausea* in Greek means seasickness (Gr. *naus*; ship).

One of the major conditions affecting the inner ear and leading to vertigo is Ménière's disease in which an excess of fluid in the inner ear causes a swelling of the tubes and cavities in the inner ear. The author of "*Gulliver's Travels*", Jonathan Swift suffered most of his life from periodic bouts of deafness, sometimes combined with illness or giddiness. He described rarely feeling safe from attacks of vertigo or from a deafening sound like rushing water in his ears that blocked out human voices. This worsened as he grew older, which, possibly in conjunction with Alzheimer's disease, led to him being declared of unsound mind three years before his death in 1742. Swift had predicted his mental decay when he was about fifty and had remarked to the poet Edward Young when they were gazing at the withered crown of a tree: "*I shall be like that tree; I shall die from the top.*" It seems likely that the disorder that affected Swift was Ménière's disease, though it was over 100 years later, in 1861, that the disorder eventually acquired a name through the French physician Prosper Ménière. Although Ménière's disease was originally assumed to be sporadic, over half of cases report other family members with the disorder, though no genes associated with the condition have yet been found. A number of other high-profile people also appear to have suffered with this condition. Martin Luther often wrote about the distresses of vertigo, suspecting Satan to be the cause. More recently, Marilyn Monroe was known to experience the vertigo and compromised hearing associated with Ménière's.

Another disorder effecting balance is the rare inherited condition, known as Joubert's Syndrome. Caused by mutations in a number of different genes, it results in underdevelopment of the cerebellum, a small part of the lower portion of the brain important for coordinating movements. While the sense of balance is maintained by the organs in the ears, in conjunction with the eyes, joints, skin, and muscles it is the brain that interprets all these cues to give us an inner sense of orientation, keeping us stable and upright. The British film-maker Andrew Kötting's documentary-style movie, "*Gallivant*" (1996), records a journey he took clockwise around the coast of Britain accompanied by his eight-year-old daughter Eden, who suffers from the disorder. Premiered to great acclaim at the Edinburgh Film Festival, it won the *Channel 4 Best New Director* prize thus raising awareness of this medical condition.

Jonathan Swift. Charles Jervas, circa 1730s.

Chapter 17

NEUROLOGICAL DISORDERS

The man who views the world at 50 the same as he did at 20 has wasted 30 years of his life.

Muhammad Ali

The nervous system has two distinct parts: the central nervous system (CNS) consisting of the brain and spinal cord, and the peripheral nervous system describing the nerves outside the brain and spinal cord. The basic unit of the nervous system is the nerve cell, also known as a neuron. These cells contain fibres called axons which send electrical impulses, and dendrites that branch out from the cell to receive these impulses. Neurons of the peripheral nervous system can be functionally classified into motor neurons and sensory neurons. Motor neurons transmit signals from the CNS to stimulate effector cells such as muscles, while sensory neurons convey information, in the opposite direction, from tissues to the CNS.

A progressive loss in coordination can result from a degeneration of neurons in the cerebellum and nerve tissue in the spinal cord, important for controlling muscle movement in the arms and legs. This can lead to the development of clumsy or awkward movements and unsteadiness, known as ataxia. The renowned British logician, Philip Edward Bertrand Jourdain, began to experience difficulties in walking as a young boy and by the time he became a student at Cambridge University he was crippled. The type of hereditary ataxia from which Jourdain suffered is Friedreich's ataxia (FA), named after the German physician Nicholaus Friedreich, who first described the condition in the 1860s. While many sufferers do not survive past their twenties, Jourdain lived just short of his 40th birthday, dying in 1919. Although rare, FA is the most prevalent inherited ataxia, affecting around 1 in 50,000 people. Inherited recessively, it usually results from the triplet repeat mutations in a gene coding for the frataxin protein involved in mitochondria function. When this is disrupted nerve cells of the spinal cord are unable to inefficiently use cellular energy leading to their degeneration. This disorder has a disproportionally high incidence among French Canadians which points to a founder effect. It seems that many inherited the mutant gene for FA from the married couple Jean Guyon and Mathurine Robin both of whom emigrated from Normandy in 1634.

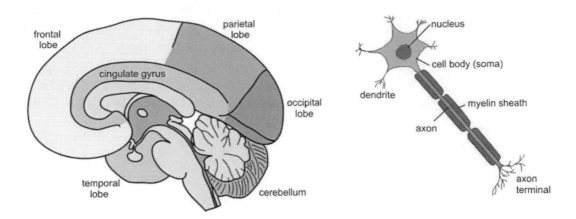

A diagram of the human brain and a neuron

Philip Edward Bertrand Jourdain.

While FA is recessively inherited, there are other forms of spinocerebellar ataxia which are dominantly inherited. These are caused by mutations in different genes encoding proteins found in nerve cells of the cerebellum and spinal cord. To date, there are 29 such genes identified, named SCA1-SCA29, most of which are affected, like FA, by triplet repeat mutations. One of these genes (SCA5) is found in the family tree of President Abraham Lincoln. Either Captain Abraham Lincoln or Bathsheba Herring must have had the affected gene and passed it on to at least two of their five children who subsequently suffered from spinocerebellar ataxia. President Lincoln would have had a 25 percent chance of inheriting the gene. As some of the other affected members of his family only developed symptoms as late as 68 years of age it could be possible that he himself had inherited the gene but symptoms had not yet manifested themselves by the time of his assassination at the age of 56.

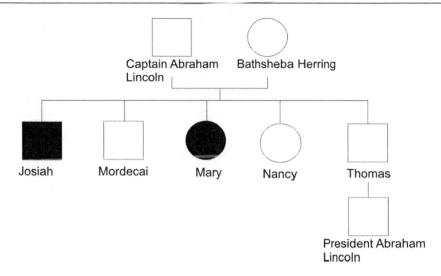

Spinocerebellar ataxia in the Lincoln family.

Neuropathies describe disorders involving the peripheral nerves. If motor nerves are damaged, muscles may weaken or become paralysed, and if sensory nerves are damaged, sensation may be lost or abnormal sensations may be felt. The most common hereditary motor and sensory neuropathy is Charcot-Marie-Tooth disease (CMTD) which affects around 1 in 2,500 people. Named after the three doctors who discovered the gene in the early 20th century, most forms of CMTD usually result from damage of the membrane surrounding neuronal axons, known as the myelin sheath, which enables nerve impulses to travel quickly. Composed of fatty substances known as lipids, myelin can be likened to the insulation around an electrical wire, acting as an insulator to electrical signals. In demyelinating diseases myelin sheaths become degraded impeding the transport of impulses which would normally travel around 225 miles an hour. In the case of CMTD this leads to a slow progressive degeneration of muscles in the feet and lower legs which can spread to the hands and forearms, and generally presents in early adulthood. Todd MacCulloch retired from playing basketball with the *Philadelphia 76ers* in 2003 after had been diagnosed with this condition which had been slowly eating away at his physical strength, starting with numbness and tingling in his feet and then spreading to his hands. He now works as a commentator. CMTD can be described as a genetically heterogeneous disorder in that mutations in different genes can produce the same disorder, resulting in different patterns of inheritance. The most common type of CMTD results from mutations in the PMP22 gene on chromosome 17. This gene produces a protein called peripheral myelin protein 22 which plays a role in the development and maintenance of myelin in the peripheral nerves.

The most common demyelinating disease is multiple sclerosis where the myelin is destroyed by the body's own immune system. First clinically described by the same French neurologist Jean-Martin Charcot, who also first reported CMTD and amyotrophic lateral sclerosis (sometimes known as sometimes called Maladie de Charcot), the resulting damage to the myelin leads to the occurrence of muscle weakness, coordination problems, and feeling of numbness or pain, with half of suffers also displaying cognitive impairments. However, the symptoms are highly variable though usually presenting between the ages of 20 and 40. This disease appears to result from an autoimmune disease where the body launches a defensive attack against its own nerve-insulating myelin affecting communications between neurons in the brain with those of spinal cord. Though it is still not properly understood there are suggestions that this might be triggered by some as yet unknown environmental trigger, possibly in combination some level of genetic susceptibility.

The film "*Lorenzo's Oil*", staring Nick Nolte and Susan Sarandon is based on the true story of Augusto and Michaela Odone, two parents in a relentless search for a cure for their son Lorenzo who was born in 1984 with an inherited demyelination disease known as adrenoleukodystrophy (ALD). This is a metabolic disorder, where the absence or malfunctioning of an enzyme results in the toxic accumulation of chemical substances damaging the nerves. The most common type is called X-linked ALD which affects around 1 in 10,000 males and results from the accumulation of high levels of Very Long Chain Fatty Acids (VLCFA) in the brain due to mutations in the ABCD1 gene that produces a protein necessary for the breakdown of VLCFA. This disease usually presents in early life when children that have previously developed normally, are quickly deprived of sight, hearing, speech and movement.

Lorenzo Odone and his family.

ALD had only been identified 10 years before Lorenzo was born so consequently relatively little was known about it. The Odones set about learning everything possible about the disease and eventually came upon a treatment that seemed to prevent the disease from progressing any further in their child. For this, Augusto Odone received an honorary PhD. Known as *Lorenzo's Oil*, the treatment is a mixture of two acids that prevents the build-up of VLCFA in the body. Outliving his expected life expectancy by two decades Lorenzo sadly died in 2008 at the age of 30 though the patent money from the oil continues to support an international research enterprise founded by the Odones called the *Myelin Project*.

Other mutations in this same ABCD1 gene result in a disorder known as Adrenomyeloneuropathy (AMN), which has an adult-onset, typically between the ages of 20 and 30. The Chilean musician Sebastian Santa Maria suffered from this. He was diagnosed with the disorder at the age of 34, while he was working on the recording of his first solo album "*Latino*", and died 3 years later in 1996. His brother Julio had also suffered and died from AMN in the 1960s, at the age of 19.

Hereditary spastic paraplegia, describes a group of inherited neurological disorders in which only the legs gradually become weak. The age at onset and the degree of muscle weakness and spasticity may be extremely variable occurring as early as infancy or as late as old age. Initial symptoms typically include stiffness and relatively mild weakness of leg muscles, balance difficulties, unexplained trips and falls, and an unusually "clumsy" gait. As the disorder progresses, walking may become increasingly difficult although complete loss of the ability to walk is relatively rare, and lifespan is not affected. Mutations of many different genes may cause hereditary spastic paraplegia, and, in most cases, these mutations appear to be transmitted as an autosomal dominant trait, though some forms show either autosomal recessive or X-linked recessive inheritance. Affecting around 1 in 30,000 people, the underlying causes are unknown though the symptoms appear to result from progressive degenerative changes of regions of the spinal cord that convey motor impulses from the brain to muscles involved in controlling certain voluntary movements.

In the 19th and early 20th centuries a diagnosis that was almost exclusively associated with women was called *neurasthenia*, or a "*weak nervous system*". These women would present with symptoms of fatigue, weakness, dizziness and fainting, and the doctor's orders would have simply been bed rest. While some of these women died, others recovered, without anyone understanding where the problems came from. Virginia Woolf was known to have been forced to undergo rest cures, which she describes in her book "*On Being Ill*". Americans were supposed to have been particularly prone to neurasthenia, which resulted in the disease acquiring the nickname of *Americanitis*. It is now thought that many of these cases were actually disorders of the autonomic nervous system, known as dysautonomia. While the neuropathies discussed so far have involved the somatic nervous system, which regulate activities under conscious control, such as body movement, dysautonomia affects the autonomic system which regulates functions not under conscious control, such as heart rate or breathing. Dysautonomia can lead to the heart rate increasing for no apparent reason, or the kidneys suddenly failing to properly retain water. Chronic fatigue syndrome is often associated with dysautonomia due to an inability of the heart and circulatory system to compensate for changes in posture, causing dizziness or fainting when one stands suddenly.

One of the most severe dysautonomic diseases is known as Ondine's Curse or congenital central hypoventilation syndrome, in which there is no autonomic control of breathing. People affected with this disorder are unable to breathe spontaneously and must consciously control each breath. Sufferers stop breathing if they ever fall sleep, triggering them to awake and take a voluntary breath again. Affecting around 1 in 200,000 births, most individuals do not survive infancy, and those that do usually require a lifetime on respirators. Ondine's Curse derives its name from the German myth of a water nymph, *Ondine*, who fell in love with a human, thereby forfeiting her immortality. Though the human had pledged his undying love, *Ondine* discovered him snoring in the arms of another woman. She cursed him, declaring that as he had pledged his love with every waking breath, he would die the moment he fell asleep. Nearly all cases of this disorder are caused by mutations in an important developmental gene which plays a role in organising the autonomic nervous system.

There are many diseases that present with dysautonomia; some are inherited, such as familial dysautonomia, while others can be acquired later in life, such as multiple system atrophy (also known as Shy-Drager syndrome). Familial dysautonomia, which is seen almost exclusively in Ashkenazi Jews and inherited in a recessive fashion, is classically characterized by an inability to produce tears, a difficulty maintaining body temperature and poor growth and muscle tone. Sufferers may also show frequent vomiting attacks and problems with speech and movement and might often hold their breath for prolonged periods of time. In 1997 Johnny Cash was diagnosed with multiple system atrophy in which neurodegeneration in specific areas of the brain impair movement, balance and autonomic functions of the body such as blood pressure, heart rate, bladder function and digestion. Despite these health issues, he still he carried on recording and performing, relying on the inspiration and support his wife, June Carter Cash. Less than four months after his wife's death Johnny Cash also died.

Johnny Cash 1969.

The distance between insanity and genius is measured only by success.

Bruce Feirstein (screenwriter) *James Bond, Tomorrow Never Dies.*

Three men once met in a resort town on a cold afternoon in February to discuss some long standing issues. One was an alcoholic with bipolar depression disorder; another was suffering from vascular dementia, while the third, long diagnosed with clinical paranoia, was also suffering from the first signs of dementia caused by a series of strokes. Not unusual, unless you consider that the town was Yalta, the year 1945, and the issues being discussed would decide the fate of post-war Europe and affect the future of the World for decades to come. One of the men, the US president Franklin D. Roosevelt was suffering the final stages of severe hypertension from which he was to die two months later. In addition he was widely suspected to have suffered from vascular dementia, resulting in signs of cognitive impairment and reduced concentration that many believe seriously impaired his negotiations with Stalin. The Soviet Union leader himself also suffered from severe psychoses and nervous breakdowns. Many millions of Russians might have had their lives spared if Stalin had been properly treated by a psychiatrist; a number of doctors who offered diagnoses, paid with their lives. History would have been very different if it were not for untreated dementias in some of the world's leaders, past, and present.

The Big Three: Winston Churchill, Franklin D. Roosevelt and Stalin at the Yalta Conference, February 1945.

After making his diagnosis of Alzheimer's disease public in 1994, former President Reagan informed the nation of his condition with his typical optimism: "*I now begin the journey that will lead me into the sunset of my life. I know that for America there will always be a bright dawn ahead. Thank you, my friends. May God always bless you.*" But it was some years earlier, while he was in his second term of office, that some psychologists began to detect possible signs of the disorder in his conversation, speech and behaviour. At one press conference President Regan was asked about his plans for talks with the Russians on space weapons. He seemed confused by the question and was unable initially to find the words. Nancy Reagan whispered loudly, "*Tell them we're doing everything we can*". "*We're doing everything we can,*" echoed the President. Remarkably, despite his growing dementia, he is remembered chiefly for his astute handling of the cold war, often against general advice. After leaving office as a hero, the disease went on to slowly ravage his mental capacity until he died of pneumonia on June 5, 2004.

President Ronald Regan, 1983.

Clinically, dementia (Lat. *de*; away; *mentis*; mind) is the progressive decline in cognitive function due to damage or disease in the brain beyond what might be expected from normal aging. Alzheimer's disease is the most common type of dementia and occurs due to the degeneration of tissue in certain areas of the brain leading to a progressive deterioration in mental functioning. This particularly affects older people; about 1 in 5 over the age of 80 suffer from it. While the causes of Alzheimer's are still not precisely known, genetic factors are implicated in familial and early-onset cases, which account for around 7 percent of all Alzheimer sufferers. The best-selling British author Terry Pratchett is suffering this form of early-onset Alzheimer's disease. Even the late onset cases though, seem to have some genetic components with epidemiological studies suggesting that there is a higher chance of both twins suffering Alzheimer's disease then there would be in the general population. Nevertheless, the only gene implicated so far is the APOE gene coding for a protein involved in the transport of lipids and in the metabolism of cholesterol, although it is still not properly understood how this leads to an increased risk.

One of the first effects of Alzheimer's disease is the gradual loss of short-term memory and the ability to reason and concentrate. The British Prime Minister, Harold Wilson, first noticed these effects while still in office, becoming aware that his short term memory was no longer working as effectively, even though his long term memory appeared relatively unaffected. It was the realisation of this, that, in part, may have prompted his unexpected resignation in 1976. He may have been influenced by the experience of seeing the mental decline of his mother who died from Alzheimer's disease.

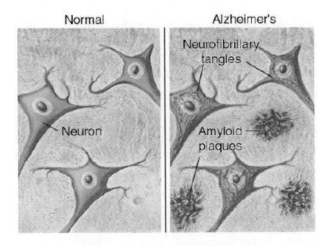

The formation of amyloid plaques and neurofibrillary tangles are thought to contribute to the degradation of the neurons in the brain.

Alzheimer's disease is just one of a number of different neurodegenerative disorders, including Parkinson's disease, that result from the deterioration of neurons in the brain. Many of these neurodegenerative diseases result from the accumulation of insoluble aggregates of different proteins building up in specific areas of the brain. In Alzheimer disease the protein aggregates are known as tau neurofibrillary tangles and amyloid plaques. The clumps of hard insoluble amyloid plaques (Lat. *Amylum*; starch: it was once thought that these masses of protein were made of starch) typically occur only outside of the neurons while the tau deposits are found inside the actual neurons. These two proteins, amyloid and tau, play important roles in the body when properly deployed and much research focuses on how and why these two proteins suddenly aggregate in Alzheimer's disease killing those neurons in certain region of the brain crucial for learning and memory.

Maurice Ravel, 1912.

Tau is not only implicated in Alzheimer's disease. The abnormal deposition of this protein in other areas of the brain leads to a range of other symptoms and diseases. The tau protein can also aggregate in the frontal and temporal lobes leading to the degeneration of this part of the brain, resulting in behavioural abnormalities rather than memory loss. This leads to symptoms including a decline in social behaviour, emotional blunting, changes in eating habits, reduced attention-span, and speech and language problems. This type of frontotemporal dementia is known as Pick's disease, after the Czechoslovakian psychiatrist Arnold Pick and accounts for around 10 percent of all cases of pre-senile dementia. Oddly, as well as negative effects, this disease has also been known to heighten creativity. The Dutch painter de Kooning, showed a dramatic change in his artistic style after developing this form of dementia, creating the expressionist abstract style for which he is so well remembered. The increase in creativity in cases of frontotemporal dementia is also exemplified by French composer Maurice Ravel who, in the last 10 years of his life, composed his most famous pieces, including *"Boléro"*, with its evocative, haunting melody. Although not displaying classical Mendelian patterns of inheritance, studies show that a significant portion of sufferers have at least one affected first-degree relative suggesting a genetic link.

Yet another disorder caused by aggregation of the tau protein, this time in the midbrain, is called progressive supranuclear palsy. One well known sufferer was the British actor and comedian Dudley Moore who inherited the mutant tau gene. The midbrain is important in balance and eye-movements, so typical symptoms involve problems with control of balance, and an inability to control movements of the eye, particularly vertical movements. This results in the so-called "dirty-tie syndrome" because those affected cannot see that they are dropping food when they eat. Dudley Moore, who was also an accomplished pianist, first noticed a problem with the co-ordination of his finger movements while playing. He was finally diagnosed when doctors found that he had the tell-tale restricted vertical eye movements which would later progress into more severe vision and balance problems and dementia. Before his death in 2002 he had set up the *Dudley Moore Research Fund*, raising awareness and funds for research in finding a cure for progressive supranuclear palsy.

Dudley Moore, 1991.

While the above neurodegenerative diseases are primarily characterised by dementia, others are associated with movement complications. The progressive movement disorder, Parkinson disease, is the second most common neurodegenerative disease after Alzheimer's disease, affecting around 1 in 500 people, with symptoms usually appearing around the ages of 50 to 60. Parkinson's disease has played a major role in events of the last century by affecting a number of former dictators. It is likely that Parkinson's disease was a key factor in Adolf Hitler's downfall. He first began to show symptoms in 1934 with newsreels showing tremors in his hand and a shuffling walk. As with so many governments, his medical condition was kept secret and by the time of the Normandy landings, he had suffered from the disease for 10 years, and had in addition developed cognitive problems suggestive of dementia. Around a third of Parkinson's suffers also develop dementia. However, Hitler would not be the last European fascist dictator to succumb to the effects of Parkinson's disease. Francisco Franco was diagnosed with the disease in the 1960s and spent his last six years of rule in a highly weakened, often bed-ridden, state. Parkinson's disease (or possibly motor neuron disease) has also been implicated in the death of the Chinese dictator, Mao Zedong in 1975. His successor Deng Xiaoping also suffered from the disease. Both men continued to rule for several years whilst being plagued with the symptoms.

Although the symptoms of Parkinson's disease have been described since ancient times, it was not formally recognized until 1817 when the British physician, James Parkinson, first documented the medical manifestations. As well as having a possible genetic link, the disease also has environmental triggers, such as the ingestion of certain toxins or repetitive head injuries and concussion. This latter may have contributed to the symptoms seen in one of the most prominent sufferers of Parkinson's disease alive today, Muhammad Ali. Symptoms of his disease began in 1982 with tremors, rigidity of muscles and slowness of speech and movement.

The effects of Parkinson's disease stem from the damage and loss of neurons in a part of the brain controlling muscle movement. These particular neurons make a chemical called dopamine which normally sends signals to coordinate movements. In a similar way to the neurodegenerative disorders previously discussed, aggregates of a protein, specifically alpha-synuclein in the case of Parkinson's disease, are generally found in the affected brain. For around 15 to 20 percent of patients there is some family history of the disease but the extent to which this is a result of shared genes rather than shared environmental risk is still uncertain. However, in the same way as Alzheimer's disease, genetic factors appear to be more predominant when the disease begins earlier in life. Several genes are reported to underlie the early onset forms of Parkinson's disease, in which development of symptoms begin at 40 years or younger. Generally inherited recessively, one prominent sufferer is Michael J. Fox who first noticed symptoms at the young age of 29. He awoke one morning in 1990 with a persistent tremor of his left little finger; this then progressed to other areas of his body. He now focuses most of his energy on his *Michael J. Fox Foundation for Parkinson's Research* which has raised much awareness and contributed many millions of dollars for research into the disorder. Michael was prescribed Levodopa, a dopamine substitute, to help control his symptoms allowing him to continue his acting career to 2000. While Levodopa treatment for Parkinson's is credited to the Swedish Nobel prize-winning scientist Arvid Carlsson, macuna seeds which contain levodopa are known to have been used in India to relieve the symptoms of palsy for at least 4,000 years.

Muhammad Ali and Michael J. Fox, during testimony before a congressional panel in favor of legislation to expand federal embryonic stem cell research, 2007.

There are a number of genes known to underlie young onset Parkinson's disease, including the previously mentioned the tau gene and the alpha-synuclein gene that also involved in Lewy body dementia. Mutations in another gene called parkin result in a disorder similar to Parkinson's, called Parkin disease. The parkin gene functions by producing the protein involved in attaching special molecules to defective proteins, tagging them for destruction - protein degradation is an important function because as proteins age, they can slowly become damaged and toxic to the cell. Though sharing many clinical features, Parkin disease is pathologically very different to Parkinson's disease. Though the same cells are affected in both cases, the brains of those with Parkin disease do not contain the typical protein deposits seen in Parkinson's disease. This disease is inherited in a recessive manner with the majority of cases developing relatively early in life, but showing a very slow progression.

Ozzy Osbourne, 2008.

Ozzy Osbourne was diagnosed with Parkin disease in 2005 claiming that he first noticed he had tremors in his 20s. *"I'd always assumed it was the booze and stuff,"* he explained, *"Now I've found it all stems from the family. When I told my sister she said, 'Not you as well? Mum had that and Auntie Elsie and your grandma.' I'm like, 'Thanks for f**king telling me'. Me walking around thinking I've got some drug paralysis."* He now takes medication to combat the involuntary tremors.

Too much sanity may be madness. And maddest of all, to see life as it is and not as it should be.

Miguel de Cervantes

As previously discussed, Charles Dickens seems to show in his novels a remarkable awareness of medical conditions. It is possible that he described the first case of obesity hypoventilation syndrome in *"The Pickwick Papers"*, and he may have been the first to appreciate the existence of dyslexia in *"Bleak House"*. It is now speculated that the character *Ebenezer Scrooge* might have been based on someone with a type of dementia known as Lewy body dementia, although it was a full century after Dickens wrote *"A Christmas Carol"* that the disease was first scientifically described. Lewy body dementia is characterised by symptoms similar to both Alzheimer's disease and Parkinson's disease but is in addition characterised by vivid and detailed hallucinations. Like Alzheimer's it is caused by accumulation of protein aggregates in the brain, but in this disease the protein aggregates are called Lewy bodies and consist of the protein alpha-synuclein.

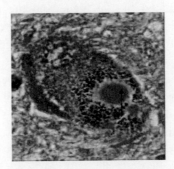

A Lewy body aggregate (pink) inside a neuron.

Almost 250 years before Dickens, the Spanish author Miguel de Cervantes Saavedra may also have incorporated the symptoms of Lewy body dementia into his writing. It is suspected that the behaviour of a patient he might have witnessed with the disorder was translated into the character *Don Quixote* in the novel of the same name. The eponymous hero suddenly becomes convinced that he is a knight errant. He dons an old suit of armour, renames himself *Don Quixote de la Mancha,* and sets off on fantastical adventures such as battling giants, which were in reality windmills. This novel ends with *Don Quixote's* complete disillusionment and a melancholy return to sanity before his demise. Two well known writers whose work appears to have been affected by their encroaching Lewy body dementia are the Prussian philosopher Immanuel Kant and the British artist, poet and playwright, Mervyn Peake. The latter's sketches during his illness portrayed the visual hallucinations he was experiencing while paranoid delusions become apparent in his poetry.

Portrait of Immanuel Kant. G. Döbler, 1791.

I am a great eater of beef, and I believe that does harm my wit.

William Shakespeare, Act 1 Scene III Twelfth Night

Prion diseases are rare infectious degenerative diseases of the brain characterised by a rapid loss of brain tissue due to the death of neurons, leaving open areas resembling a sponge – hence they are often known as transmissible spongiform encephalopathies. The death of this brain tissue leads to dementia. These diseases are thought to be caused by a protein called PrPc that normally functions in cell signalling. This particular protein however, has the capacity to convert into an abnormal form called a prion (an acronym for proteinaceous infection). These can bind together to form aggregates that destroy the cell in which it accumalates. These diseases can either be inherited, such as familial Creutzfeldt-Jakob disease (CJD), or acquired as in the variant CJD and kuru.

Mutations in the gene for PrPc can increase the likelihood of PrPc proteins converting to prions. Mutations in different areas of the gene lead to a number of different familial prion diseases that are autosomal dominant inherited with overlapping symptoms such as familial CJD, Gerstmann-Sträussler-Scheinker syndrome and fatal familial insomnia that is characterised by the degeneration of an area which influences sleep. The symptoms of CJD are a rapid progressive dementia and loss of coordinated movement, with death often occurring within six months of onset. George Balanchine, one of the 20th century's greatest choreographers, died of CJD, which was diagnosed after his death. The first signs appeared in 1978 when he began losing his balance while dancing, followed soon after by deterioration in eyesight and hearing, until he became totally incapacitated, dying at the age of 79. Around one in every million people each year develops CJD, although this may be a low estimate; the disease is very difficult to diagnose as there is no definitive diagnosis without examining the brain.

Only a small percentage of prion disease cases run in families with most prion diseases developing sporadically in the absence of any mutation or as a result of infection with abnormally formed prion protein. This was first noted in the Fore tribe of Papua New Guinea in the early 1900s. During funeral rituals members of the tribe would eat the brains of dead relatives leading to the spread of a prion disease known as kuru, which means "*trembling*" in the language of the Fore. In a similar way, some cases of CJD in the U.S. have developed from the use of human growth hormone extracted from the pituitary glands of patients who died from CJD, though the known incidence of these cases is small and this procedure was discontinued in 1977. It is possible that William Shakespeare might have provided the first account of CJD, in his character of *Macbeth* who experienced a rapid descent into madness and a decline in neurological and cognitive function accompanied with involuntary movements, hallucinations and insomnia. It is conceivable that he might have even had an understanding of the transmission of the disease by consumption of infected neuronal tissue: "*...when the brains were out, the man would die. And there an end; but now they rise again. With twenty mortal murders on their crowns*." For Macbeth this might have been from the consumption of the brew given to him by the *Weird Sisters*, which contained various parts of humans and animals.

Suzanne Farrell and George Balanchine dancing in "*Don Quixote*" 1965

Prion diseases occur in many other animals such as cows, where it is known as mad cow disease (bovine spongiform encephalopathy), sheep (scrapie), cats, and even ostriches. However different animals have different amino acid sequences for the PrPc protein and generally only PrPc protein molecules that are identical in amino acid sequence to the prion protein can be recruited into the growing fibre and aggregate. This 'specificity' phenomenon is thought to explain why transmission of prion diseases from one species to another (such as from sheep to cows or from cows to humans) is likely to be a rare, if not impossible, event. However, a very small number of people in the UK may have contracted a variant form of CJD from eating infected beef.

STEM CELLS

Stem cells are undifferentiated cells that retain the ability to divide and differentiate into many other cell types. These cells are found naturally in certain areas of the adult body, and in the embryo, where they serve to replenish cells in the body. The main types of stem cells are the embryonic stem cells and adult stem cells which differ in their "potency". Adult stem cells are described as "multipotent" meaning that they can develop into closely related cells though are unable to differentiate into every cell type in the human body. For example, hematopoietic stem cells can only differentiate into red blood cells, white blood cells and platelets. Embryonic stem cells, on the other hand, are described as being pluripotent in that they can form any of the 200 cells of the adult human body.

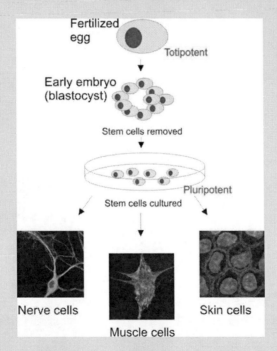

Embryonic stem cells are cells derived from the inner cell mass of developing blastocysts. An ES cell is self-renewing (can replicate itself), pluripotent (can form all cell types found in the body) and theoretically is immortal.

These embryonic stem cells are found only in the very early stage embryo which, in humans, are approximately 4 to 5 days old and consist of 50 to 150 cells. Because of their unique combined abilities of unlimited expansion and pluripotency, embryonic stem cells are a potential resource for regenerative medicine and tissue replacement after injury or disease. Medical researchers anticipate being able to use technologies derived from stem cell research to treat cancer, Parkinson's disease, spinal cord injuries, and muscle damage, amongst a number of other diseases, impairments and conditions. However, there is widespread controversy over stem cell research relating mainly to the way in which stem cells are acquired from human embryos resulting in the destruction of a possible human life. Although President George W. Bush became the first president to provide federal funding for embryonic stem cell research, totalling approximately $100 Million, in July 2006 he vetoed a bill that would have allowed this money to be used for research which involves the destruction of embryos. New research, however, seems to suggest that it may be possible to alter certain genes within a normal non-stem cell allowing a differentiated cell to regain its pluripotency.

Chapter 16

CANCER AND DISORDERS OF DNA REPAIR

When you think about it, what other choice is there but to hope?
We have two options, medically and emotionally: give up, or fight like hell.

Lance Armstrong

Hippocrates was one of the first to describe different kinds of cancers, using the word *oncos* (Greek for "swelling") to describe benign tumours, and *carcinos* (Greek for "crab") for malignant tumours possibly due to the appearance of the cut surface of some solid malignant tumours which vaguely resembles the silhouette of a crab. However, he was certainly not the first to discover the disease and the documentary history of cancer stretches back to 1500 BC in ancient Egypt where a papyrus describes how breast cancers were treated by using a hot piece of metal to burn away the tissue.

Cancer basically describes an uncontrolled division of cells in which malignant cells can either invade adjacent tissues or else implant themselves into distant sites (known as metastasis). This unregulated cell growth is caused by damage to DNA, resulting in mutations to specific genes coding for proteins regulating cell growth and division, known as proto-oncogenes and tumour suppressor genes. Proto-oncogenes promote cell growth in a variety of ways by coding for proteins such as hormones, signal receptors, or transcription factor proteins which can increase expression of other genes. Tumour suppressor genes, on the other hand, code for anti-proliferation signals and proteins that suppress mitosis and cell growth. These tumour suppressors are normally activated by cell stress or DNA damage, thereby serving to halt cell division in order to carry out DNA repair, preventing mutations from being passed on to daughter cells. One example of a tumour suppressor is the p53 protein, which is activated by many cellular stressors including low oxygen and UV radiation damage. A mutation damaging the gene for p53, or in a gene controlling p53, can result in a lack of this protein with the consequence that DNA repair is inhibited, allowing further DNA damage to accumulate, inevitably leading to cancer. However, mutations in several of these genes, resulting in a loss of a tumour suppressor protein or in an increasing level of a proto-oncogene protein, are required before a normal cell transforms into a cancer cell; a mutation limited to only one gene would be suppressed by the control of normally functioning tumour suppressor genes.

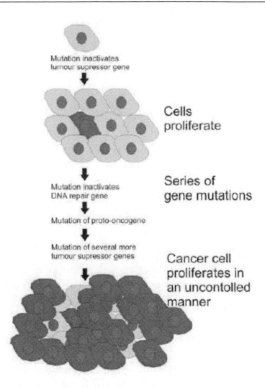

Mutations in tumour suppressor and proto-oncogenes in a cell can lead to uncontrolled cell division and growth known as cancer.

Familial cancer syndromes result from the inheritance of a mutation in a tumour suppressor or proto-oncogene gene, increasing the chance of developing particular cancers. Mutations in tumour suppressor genes generally act in a recessive manner with a mutation in the second, normally functioning, copy of the gene needed before a cell transforms into a cancer cell. For instance, individuals who are heterozygous for p53 mutations are often victims of Li-Fraumeni syndrome. As long as one copy of the p53 gene produces a functional protein, tumours do not form. However, if a mutation occurs in the second copy of the gene during a person's lifetime, the cell will have no working copies of the gene and will produce no functional protein allowing tumours to develop. Li-Fraumeni syndrome is characterisewd by an increased chance of cancers in a variety of different tissues such as breast and brain if cells here acquire mutations in the second copy of the p53 gene. In the same way, mutations in the adenomatous polyposis coli gene are linked to colon cancer, mutations in the retinoblastoma gene increase the chance of developing cancer of the retina and mutations in BRCA1 and BRCA2 can lead to early onset of breast cancer. One disease, known as Von Hippel-Lindau syndrome occurs due to mutations of the Von Hippel-Lindau tumour suppressor gene resulting in the growth of tumours in various parts of the body such as in the central nervous system, and particularly in the adrenal gland. It has been suggested that the hostility underlying the famous American Hatfield-McCoy feud of the late 1800s may have been partly due to the consequences of Von Hippel-Lindau disease. Generations of the two families fought often deadly battles over land, timber rights and even a pig and are the subject of dozens of books songs and countless jokes. It seems that the McCoy family was pre-disposed to bad tempers partly because many of them had adrenal gland tumours which even now many of their descendents still suffer from. This tumour leads to increased production of adrenaline causing a tendency toward explosive tempers in addition to high blood pressure, pounding headaches, heart palpitations, facial flushing, nausea and vomiting.

Randolf McCoy, circa 1900.

Of all cases of cancer, only a small percent derive from a hereditary predisposition, with most cancers occurring sporadically and being linked to whole range of factors including environment and life style. Therefore, the presence of one or two cases of cancer in a family does not necessarily mean an increased genetic risk. Many factors can help determine if the cancer in a family is hereditary, including the presence of certain types of cancer occurring in the same family (for example breast cancer and ovarian cancer, or colon cancer and uterine cancer), the age of onset of cancer (particularly for breast cancer and colon cancer, onset before age 50 is considered more significant), and the number of relatives with cancer and their relationship to each other.

Some theories suggest that there is a decline in the efficiency of DNA repair processes during aging, which may explain why cancer is so much more common among older people.

DNA repair stands as the dike between us and the inundation of mutations.

Robert Weinberg

An average cell normally acquires around 50,000 to 500,000 DNA defects per day. This damage occurs during normal metabolic processes inside the cell, but can also occur upon exposure to other sources of DNA damaging agents such as UV light. Whilst this constitutes only a tiny fraction of the human genome, a single unrepaired lesion to a critical cancer-related gene, as just described, can have drastic consequences. Therefore, DNA repair is an important process which constantly operates in cells. If the rate of DNA damage exceeds the capacity of the cell to repair it, the accumulation of errors can overwhelm the cell and result in cancer or cell death. Therefore, inherited diseases associated with faulty DNA repair result in increased sensitivity to agents that can damage DNA. Diseases of this kind include xeroderma pigmentosum and premature aging. Xeroderma pigmentosum (XP) was the first disorder shown to be due to defects in DNA repair – specifically, an inability to repair UV damage to DNA, leading to the development of multiple skin cancers. Normally, DNA damage from UV light is repaired through a pathway involving the cutting out of the damaged segment of DNA and replacing it with the correct sequence from the opposing strand – a process termed nucleotide excision repair. However in XP, certain enzymes needed in this pathway are missing leading to DNA damage not being readily fixed. There are actually eight types of XP resulting from mutations of different genes that alter enzymes in this process. Diagnosis can usually be made by exposing cell cultures of affected individuals with the radioactive nucleotide base thymidine after being exposed to UV light. If the radioactive thymidine is not incorporated into the DNA then this is indicative of their inability to repair the DNA damaged formed by UV exposure. Naturally, the most important part of managing the condition is reducing exposure to the sun. An episode of an American series called "*Extreme Makeover*" involved the Pope family with a daughter Shelly who suffers from XP; after the show Disney World had a special night-time opening (midnight to 4 am) so that XP children could spend time at the amusement park.

Another disorder of DNA repair is the autosomal recessive Bloom syndrome, which is due to a defect in a DNA ligase enzyme that is also involved in nucleotide excision repair. Resulting in skin lesions in response to UV exposure in addition to small stature and higher susceptibility to cancer, this disease is more frequent in Ashkenazi Jews where around 1 in 100 are carriers of the gene mutation. Further disorders of DNA repair include the previously mentioned Fanconi anaemia, ataxia telangiectasia, and Cockayne syndrome, all of which result from mutations in genes coding for various enzymes of DNA damage recognition or DNA repair.

Another disease resulting from a gene mutation leading to the disruption of a protein involved in the maintenance and repair of DNA is Werner syndrome. This is named after Otto Werner, a German student who described the syndrome as part of his doctoral thesis in 1904. Affecting around 1 in 1,000,000 individuals, this disease is characterized by accelerated aging with the dramatic and rapid appearance of features associated with normal aging. It is suggested that the altered Werner protein either interferes with the ability of cells to divide normally or that it may allow DNA damage to accumulate impairing normal cell activities. Individuals with this disorder typically grow and develop normally until they reach puberty after which a premature aging occurs involving greying and loss of hair, a hoarse voice, wrinkled skin, and the development of many disorders associated with aging such as cataracts, skin ulcers, type 2 diabetes, atherosclerosis, osteoporosis and some types of cancer.

More severe forms of accelerated aging are called progeria (Gr. *pro*; early, *geras*; old age). The classic type of this disease does not result from defective DNA repair, but occurs due to mutations in a gene for the protein lamin A which is a component of the nucleus membrane in the cell, though, how this leads to the characteristic features of progeria is still not known. It has been suggested that the American author F. Scott Fitzgerald could have been describing this disorder in his short novel "*The curious case of Benjamin Button*" published in 1921 about a child born with features of an elderly man. First identified in 1886 by Jonathan Hutchinson and Hastings Gilford, this extremely rare genetic condition, affecting around 1 in 8 million newborns, leads to characteristics of accelerated aging occurring at around 18-24 months of age with most sufferers dying in their early teens from heart disease or other age-related problems. While the disease can be dominantly inherited most cases appear to be due to a mutation occurring during the early stages of cell division in a newly conceived child or in the gametes of one of the parents, i.e. a de novo mutation. "*The Child*

Who's Older Than Her Grandmother" is a UK documentary first screened in 2005, telling the tragic story of Hayley Okines, a six year old girl born with progeria. However, despite this terrible disease, she became a happy and popular child at school and put a lot of energy into raising the awareness of progeria including recording a song with the *Kids Choir 2000 in which*; all proceeds went to the Progeria Research Foundation.

Hayley Okines.

Around 30 years ago, a middle-aged lawyer in France struck a deal with an old lady in her 90s: he gained ownership of her apartment after death, in return for a large monthly allowance. It seemed on the face of it a mutually beneficial proposition. To the lawyer's great misfortune, his client, Jeanne Calment, was destined to become the longest lived person in modern history. She died in 1997, at the age of 122 years and 164 days, with all faculties intact; when asked on her 120th birthday what kind of future she would expect to have, she quickly replied "*A very short one.*" Unfortunately her lawyer died long before she did, and his wife was forced to carry on the payments as part of the deal - paying Jeanne Calment the price of her apartment many times over. Calment's ancestors going back at least five generations on both sides of her family had lived, on average, a remarkable 10.5 years longer than the mean age at death of people in the same region, leading one to conclude that how she lived – she smoked until the age of 117 - was not the main factor in her great age. Over 100 years ago, Alexander Graham Bell investigated the inheritance of longevity and he came to the conclusion that marriage partners, who both come from families characterised by longevity, are more likely to have long lived offspring. This supposition was confirmed many decades later from experiments with fruit flies showing it was possible, by mating flies produced from the eggs of long lived flies, to produce a breed of flies living 30 percent longer than normal, after only 10 generations. However, these long-lived flies proved to act more sluggishly than flies with a normal lifespan; they also had a lower metabolic rate and, interestingly, had decreased fertility.

There are a number of hypotheses concerning how aging occurs. Oxygen radicals, which are by-products of normal metabolism, are often considered as the villains of the cell, damaging proteins, membranes, and DNA. The body has a defence system to disarm these molecules, involving some vitamins and enzymes such as superoxide dismutase. However, little by little the damage mounts up contributing to deteriorating organs and tissues. It is, therefore, interesting that a variety of fruit fly, containing a more active version of the superoxide dismutase gene, lives up to 40 percent longer. Other suspects in cellular deterioration are known as advanced glycosylation end products (AGEs), resulting from glucose attaching to proteins in the cell and causing them to crosslink with each other. Although this process is slow, it does increase with time and is thought to result in some of the deterioration associated with aging, such as a stiffening of connective tissues, hardened arteries, cataracts, and loss of nerve function. Again, the body has its own defence mechanism against this destruction which may differ among individuals. As previously discussed, defects in DNA repair can lead to a susceptibility to cancer and accelerated aging.

Perhaps related to all of the above theories are the findings of research linking calorie intake to longevity. It was first observed in 1934 that rats fed on a severely reduced calorie intake can have lifespan of up to twice as long as control rats. These findings have been successfully repeated in a number of other animals, though research in primates and humans are still ongoing. Nonetheless many people have independently adopted the practice of a calorie-restricted diet, in the hope of achieving a longer life.

Another recent discovery is a built-in limitation on cell division; cells after a certain number of divisions enter a state of senescence whereby they do not divide or proliferate further. For example, human fibroblast cells divide about 50 times and then stop. Moreover, fibroblasts taken from an elderly person have fewer divisions left then than those of a child. One mechanism underlying this process involves special DNA tails at the ends of chromosomes, called telomeres, which get shorter as the cell divides. This apparent counting mechanism, keeping a track of the cells age, may regulate cell lifespan in some way. Indeed, a number of the premature aging syndromes, mentioned earlier in the chapter are associated with short telomeres and furthermore, in cancer cells that have become immortal (i.e. the cells divide indefinitely) these telomeres stop shrinking with each cell division. This has led to the discovery of an enzyme called telomerase which, in cancer cells, replaces the lost telomere sequences on the ends of the chromosomes. A number of drugs are being designed to target the telomerase enzyme, in an attempt to control cancer, and research is ongoing to determine if manipulation of this enzyme may have an effect on human aging.

Aurora, Roman goddess of the dawn, bids goodbye to her lover Tithonus. Francesco Solimena, 1704.

Concluding, one should take a lesson from Greek mythology, and the story of *Eos* the goddess of the dawn. She had a lover called *Tithonus*, who was mortal, and she asked *Zeus* for one wish - for *Tithonus* to be made immortal. However, she forgot to ask for *Zeus* to also give him eternal youth. So while at first, they were happy together, he inevitably began to grow old, while she remained young and beautiful. *Eos* soon grew tired of *Tithonus*. As he aged he became increasingly debilitated and demented, driving *Eos* to distraction with his constant jabbering. Eventually, *Eos* locked *Tithonus* away in a room where he presumably still lies becoming ever more decrepit.

GLOSSARY

Allele One of alternative forms of a gene at a particular locus.

Amino Acid One of the building blocks of proteins.

Anticipation The process whereby some genetic diseases get more severe in each successive generation.

Autosome A chromosome which is not a sex chromosome.

Barr body An X chromosome which is condensed and inactive and which shows as a dark staining blob in an interphase nucleus.

Base A basic part of a nucleic acid. Adenine (A), thymine (T), cytosine (C), guanine (G) and uracil (U). *See base pair, nucleoside and nucleotide.*

Base pair (bp) The fundamental unit of a double stranded DNA molecule, (more strictly - a nucleotide pair). The base Adenine on one strand is paired with Thymine on the other and Guanine with Cytosine. The lengths of double stranded DNA molecules are frequently given in bp (or nucleotide pairs).

Carrier A person heterozygous for a recessive trait.

Chromosome The structure which is built up around each nuclear DNA molecule. It is comprised of a single (double stranded) DNA molecule with associated histone proteins, non-histone proteins and RNA. It is most easily viewed at its most condensed (at metaphase of mitosis) when its structure has been duplicated to give two chromatids. All normal human chromosomes have a centromere, two arms and two telomeres.

Clone Noun; frequently used to mean a DNA molecule which has been replicated in a micro-organism such as a bacterium or yeast to make many thousands or millions of identical molecules. It also means an individual who is genetically identical to another individual, which might mean an identical twin.

Verb; to carry out the process of replicating an individual or a specific DNA molecule.

Coding sequence That part of a gene which contains the codons which will, via a mRNA intermediate, be translated into polypeptide.

Codon A group of three consecutive nucleotides in mRNA which specifies an amino acid to be incorporated in the polypeptide product of the gene. *See genetic code.*

Congenital Present at birth but not necessarily inherited.

Consanguineous mating A mating in which male and female are related by descent, i.e inbreeding.

Deletion A mutation resulting in the loss of normal DNA sequence. A deletion may be of any size from 1 nucleotide pair to the loss of a piece of chromosome.

Deoxyribonucleic acid *see DNA*

Dizygotic twins Twins arising by the fertilisation of two eggs by two sperm.

DNA The molecule in which the genetic information of most organisms is encoded.

DNA fingerprinting A process by which an individual can be uniquely identified by testing for multiple DNA polymorphisms.

Dominant An allele is dominant if its effect can be observed in the phenotype of a heterozygote.

Epigenetic Any factor that influences the phenotype which is not part of the genotype.

Eukaryote A class of organisms (which may be single or multicellular) in which the genetic material is contained within a nucleus. Note that there are further specialised structures which are also unique to eukaryotes.

Familial trait A trait which is more common in the relatives of an affected person.

Founder effect A high frequency of a particular allele in a population caused by it having been present in one or more members of a small number of individuals from whom the population is descended.

Frameshift mutation A mutation which, by deletion or addition of a number of basepairs which is not divisible by three, causes an alteration in the reading frame of a gene.

G bands or Giemsa banding The pattern of dark and light coloured Giemsa stained bands on metaphase chromosomes.

Gamete A sperm or an ovum.

Gene In Mendelian terms - a unit of inheritance. In molecular terms - a region of DNA which contains the information to create either a functional RNA or a polypeptide chain.

Gene family A number of genes which resemble each other in DNA sequence, presumably because they have evolved by gene duplication and subsequent divergence.

Gene pool All the alleles at a particular locus present in a population.

Genetic disease A disorder, which may or may not be apparent at birth, which is a consequence of a mutation present in one or more of the patient's genes.

Genetic Drift Random changes in allele frequencies from one generation to another in small populations

Genetic heterogeneity Similar phenotypes caused by mutations in more than one gene.

Genetic marker A polymorphic locus which can be used in linkage studies. The polymorphism may be anything to do with the DNA at the locus or its possible product so long as it can be recognised with an appropriate test.

Genome The complete DNA sequence of an organism.

Genotype The alleles present in an individual at the locus under consideration. Alternatively, the sum of all the alleles present in a genome.

Germline The cells which are in direct mitotic line of descent from the zygote to its gametes. As distinct from somatic cells.

Heterozygote (adj. heterozygous) An individual having two different alleles at a locus.

Imprinting The differential expression of genes depending on whether they were inherited maternally or paternally.

Insertion mutation A mutation caused by the addition of genetic material into a gene.

Inversion A chromosomal rearrangement in which a segment of chromosome is turned end on end.

Karyotype The chromosomal constitution of an individual.

kilobase (kb) 1,000 nucleotides or nucleotide pairs.

Kindred An extended pedigree.

Linkage The tendency of genes close together on the same chromosome to be inherited together. It can be quantified and used as a mapping tool.

Linkage disequilibrium The occurrence, on the same chromosome, of some combinations of alleles of closely linked genes more often than would be predicted by chance.

Locus A point on a chromosome at which a specific gene or other marker is found.

Meiosis The sequence of two cell divisions which turns a diploid germ cell into a haploid gamete.

Messenger RNA (mRNA) RNA which leaves the nucleus carrying information which, by translation, will direct the synthesis of polypeptides.

Missense mutation A mutation which changes the amino acid incorporated into the coded polypeptide chain.

Mitochondrial DNA The DNA of the mitochondrial genome.

Mitochondrial inheritance Inheritance of a character encoded in the mitochondrial genome.

Mitosis A normal cell division resulting in two genetically identical daughter cells.

Monosomy having only one copy of one of the chromosomes.

Monozygotic twins Two individuals derived from a single fertilised egg and therefore genetically identical.

Mosaic An individual composed of more than one genetically distinguishable cell population derived from a single zygote.

Multifactorial inheritance The occurrence of a phenotype as a result of the action of more than one gene and/or of environmental factors.

Mutation A change in DNA sequence from one generation (or cell generation) to the next.

Nondisjunction The failure of a chromosome pair to separate at the first meiotic division, or for two chromatids of a replicated chromosome to separate at mitosis (or at the second meiotic division), so that both pass to the same daughter cell.

Nucleotide A building block of a nucleic acid consisting of a base (adenine, thymine, cytosine, guanine, uracil) joined to a sugar (ribose or deoxyribose) and a phosphate.

Oncogene A dominant gene whose expression leads to uncontrolled cellular proliferation.

Pedigree The members of a family, also used to describe a diagram of their relationships one to another and with information on the inheritance of one or more conditions or genetic loci.

Penetrance The frequency with which individuals with the necessary genotype express symptoms of a genetic condition.

Phenotype An individual's outward appearance.

Point mutation A mutation affecting a single nucleotide pair.

Polymerase chain reaction (PCR) A technique by which a relatively small piece of DNA of known sequence can be amplified (often from a complex mixture) by successive cycles of strand separation followed by DNA synthesis (using a DNA polymerase purified from a thermophilic bacterium) primed by artificially synthesised oligonucleotide primers (one for each strand).

Polymorphism Genetic variation occurring in a population so that at least two alleles are present at a frequency of 1 percent or greater. The variation may range from alteration of noncoding DNA sequence without any phenotypic effect through to variations which gives rise to a visible change in the phenotype.

Polypeptide chain The chain of amino acids joined by peptide bonds which is the primary product of the translation of the mRNA of a gene.

Probe A DNA or RNA molecule which has been labelled and which may be used to identify its complementary sequence by hybridisation.

Prokaryote A simpler organism than a eukaryote having no nucleus and being different in many other ways too. e.g. a bacterium.

Promoter The DNA sequences (often at the beginning of a gene) that are responsible for regulating transcription and expression of the gene.

Proto-oncogene A gene which may mutate to become an oncogene.

Reading frame The way in which nucleotides are read in groups of three (codons) to specify the polypeptide coded by a gene. An RNA molecule has three possible reading frames - usually the first AUG codon defines the beginning of the reading frame selected by the ribosomes.

Recessive A mutation or allele which does not affect the phenotype unless it is homozygous.

Recombinant DNA DNA containing an artificial combination of pieces which are not found together in nature.

Retrovirus A class of viruses whose infectious genome is single stranded RNA and which, as part of their normal life cycle, integrate into their host genome after conversion of the RNA to DNA by the enzyme reverse transcriptase.

RNA The class of molecules which are the primary products of genes. One type of RNA, mRNA carries the information coded in a gene from the nucleus to the cytoplasm where it is translated into protein.

Sex Chromosome One of the chromosomes which are present in different numbers in males and females. In humans this means the X and Y chromosomes, XX in females and XY in males.

Sex-linked A trait which is caused by mutation of a gene on the X chromosome, usually called X linkage to avoid confusion with Y linked traits.

Sib A brother or sister.

Somatic cell A cell which is not on the lineage from which gametes are made. *See germline.*

Sporadic A case of genetic disease caused by a new mutation.

Syndrome A collection of symptoms which occur together and are thought to be caused by the same mutation or chromosomal anomaly.

Telomere The DNA structure which stabilises the ends of chromosomes.

Transcription The process of copying DNA into RNA mediated by the enzyme RNA polymerase.

Transgenic animal An animal into the genome of which a foreign gene has been introduced.

Translation The process of copying a mRNA into a polypeptide chain mediated by the ribosomes.

Translocation A mutation which has moved one segment of a chromosome to a different position in the genome.

Trisomy having three copies of a single chromosome

Tumour suppresser gene A gene which if inactivated by mutation allows uncontrolled cellular proliferation. Contrast with oncogene.

X chromosome A chromosome which is present in two copies in the genome of normal human females but in only one copy in normal males.

X inactivation The random inactivation of all but one X chromosome in most cells. *See Barr body.*

X linked A gene which is present on the X chromosome.

Y chromosome A small chromosome which, in humans, is present in one copy in males only.

Zygote The diploid cell formed by the fusion of two haploid gametes, i.e. a fertilised egg.

INDEX

A

abdomen, 185

abdominal cramps, 175

abnormalities, 13, 14, 15, 17, 18, 20, 51, 91, 111, 120, 122, 139, 209, 224

abortion, 36, 60, 139

accounting, 120, 122, 208

achondroplasia, 14, 52, 53

acid, 2, 6, 7, 98, 111, 134, 176, 177, 178, 180, 231, 245, 246, 247, 248

acidosis, 183

acne, 20, 169

acquired immunity, 148

acromegaly, 161

activation, 47, 145

acute, 83, 134, 136, 144, 205

acute intermittent porphyria, 134

adaptation, 1

adenine, 4, 248

adhesion, 144

adjustment, 147

administration, 17, 136

adolescence, 20, 170, 196

adrenal gland, 157, 160, 165, 167, 168, 237

adrenal glands, 157, 160, 165, 167, 168

adrenal hyperplasia, 171

adrenaline, 237

adrenocorticotropic hormone, 160

adult, 42, 77, 85, 124, 134, 159, 217, 232

adult stem cells, 232

adulthood, 51, 83, 161, 196, 203, 216

adults, 20, 124, 207

aerobic, 3

agammaglobulinemia, 149

agents, 86, 149, 151, 238

AGEs, 242

aggregates, 223, 226, 228, 230

aggregation, 224

aggression, 168

aging, 222, 238, 239, 240, 241, 242

aiding, 115

AIDS, 29

AIP, 134, 136

air, 85, 101, 107, 110, 118, 150, 170, 177, 194, 247

Albert Einstein, 211

albinism, 97, 98, 99, 100, 112, 195

Albino, 100

alcohol, 64, 134, 186

alcohol abuse, 64

alcohol consumption, 186

alcoholism, 186, 190

aldosterone, 165

aliens, 67

allele, 30, 31, 33, 151, 193, 246, 248

alleles, 17, 27, 30, 31, 33, 153, 246, 247, 248

alopecia, 106, 107

alopecia areata, 107

alpha, 112, 118, 130, 170, 171, 187, 226, 227, 228

alpha-1-antitrypsin, 118, 187

alternative, 245

Alzheimer disease, 223

amaurosis, 198

amino, 2, 6, 7, 98, 111, 134, 174, 176, 177, 178, 180, 231, 245, 247, 248

amino acid, 2, 6, 7, 98, 111, 174, 176, 177, 178, 180, 231, 245, 247, 248

amino acids, 2, 6, 7, 174, 176, 178, 180, 248

ammonia, 180

amniocentesis, 139

amniotic, 139

amniotic fluid, 139

amputation, 48

amputees, 61

amyloid, 223

amyloid plaques, 223

amyotrophic lateral sclerosis, 216

anaemia, 31, 128, 129, 130, 131, 132, 135, 138, 141, 151, 239

Anaemia, 130
Andes, 162
androgen, 20, 106, 169, 170
androgen receptors, 106
androgens, 106, 169
anhidrosis, 95
animals, 11, 46, 108, 149, 176, 231, 242
anorexia, 106
antibiotic, 149
antibiotics, 190
antibodies, 147
antibody, 149, 155
anticoagulant, 145
antigen, 148
antigens, 153
aorta, 73
apartheid, 167
APOE, 222
appetite, 112, 165, 183
arginine, 111
arrest, 185
arteries, 37, 71, 120, 121, 185, 242
artery, 144
arthritis, 90, 161, 180, 189
artistic, 64, 78, 224
assassination, 215
assignment, 18, 42
assumptions, 193
astigmatism, 196
asylum, 110
asymptomatic, 131
ataxia, 40, 213, 214, 215, 239
atherosclerosis, 240
athletes, 19, 73, 80, 120, 124, 125, 142
athleticism, 141
atmosphere, 2
ATP, 174
atria, 120
atrophy, 83, 203, 219
attacks, 118, 128, 134, 141, 147, 154, 155, 166,
    180, 197, 211, 219
auditory nerve, 92
Aurora, 243
autoimmune, 90, 107, 153, 156, 166, 183, 210,
    216
autoimmune disease, 107, 153, 156, 216
autoimmune diseases, 153, 156
autonomic nervous system, 218, 219
autopsy, 73, 117, 119, 124, 144, 161, 185, 207
autosomal dominant, 42, 53, 54, 55, 62, 71, 72,
    78, 88, 92, 110, 121, 134, 136, 179, 184, 194,
    209, 218, 230

autosomal recessive, 35, 54, 64, 66, 71, 83, 111,
    116, 119, 128, 130, 135, 175, 176, 187, 200,
    218, 239
autosomes, 12, 37
averaging, 21
aviation, 202
awareness, vii, 83, 88, 92, 155, 212, 225, 226, 228,
    240
axons, 155, 213, 215

B

B cell, 148
B cells, 148
babies, 21, 120, 123, 129, 139, 149, 157, 191, 193
back pain, 167
bacteria, 1, 2, 3, 7, 38, 42, 67, 115, 118, 149, 151,
    179, 181
bacterial, 115, 124, 151, 152, 190
bacterial infection, 124, 152
bacterium, 245, 248
balance organs, 211
barley, 191
Barr body, 16, 245, 249
barrier, 86, 118
barriers, 183
base pair, 4, 5, 11, 42, 122, 245
basement membrane, 184
basketball, 73, 120, 216
B-cell, 24, 147, 149
B-cell lymphoma, 24
B-cells, 147
beating, 64
beef, 229, 232
beliefs, 48
bell, 92
benign, 92, 160, 168, 235
bible, 103, 161
bicarbonate, 183
Big Three, 221
bile, 186
bilirubin, 186
binding, 112, 144, 157, 170
biological processes, 78, 147
biosafety, 7
biotechnology
bipolar, 190, 220
bipolar disorder, 190
birth, 13, 16, 19, 21, 29, 31, 37, 45, 48, 51, 59, 87,
    93, 97, 98, 109, 112, 120, 123, 129, 139, 140,
    150, 164, 175, 193, 198, 199, 210, 246
birth weight, 48
births, 14, 15, 16, 17, 44, 45, 56, 76, 116, 122,
    141, 158, 177, 178, 180, 182, 219

bladder, 156, 219

blaming, 94

bleeding, 99, 142, 143, 145

bleeding time, 144

blind spot, 200

blindness, 42, 65, 151, 179, 181, 198, 199, 200,
    201, 202, 203, 204, 205, 210

blocks, 2, 6, 176, 245

blood clot, 99, 121, 142, 143, 145

blood flow, 59, 120, 121, 123, 145

blood glucose, 131, 159, 165

blood pressure, 95, 144, 166, 185, 219, 237

blood transfusion, 143

blood vessels, 21, 56, 73, 85, 120, 144, 156

bloodstream, 151, 157, 160

B-lymphocytes, 147

body fat, 168

body temperature, 95, 219

boiling, 55

bonds, 36, 248

bone density, 62

bone form, 51, 64, 66

bone growth, 18, 51, 53, 54, 62, 64, 66, 93

bone marrow, 127, 138, 141, 150

bone marrow transplant, 150

bovine, 231

bovine spongiform encephalopathy, 231

bowel, 190, 191

boys, 139, 142, 163, 171

brain damage, 176

brain development, 93, 176

brain stem, 81

breakdown, 110, 181, 186, 217

breast cancer, 235, 237, 238

breast milk, 175

breastfeeding, 163

breaststroke, 125

breathing, 76, 115, 116, 165, 218, 219

breeding, 29, 37, 57, 80

brittle hair, 111

bronchitis, 119

brothers, 40, 98, 142, 180, 210

bubble, 88, 150

building blocks, 2, 6, 176, 245

Burkitt lymphoma, 24

burn, 235

burning, 95, 135

burns, 95

butterfly, 156

by-products, 178, 241

cadaver, 161, 171

calcium, 51, 65, 183

calluses, 86

calorie, 242

Cameroon, 124

cancer, 4, 88, 96, 102, 104, 133, 138, 151, 198,
    233, 235, 236, 238, 239, 240, 242

cancer cells, 242

candidates, 179

carbohydrate, 174, 186

carbohydrate metabolism, 174, 186

carbohydrates, 174

carbon, 2, 115

carbon dioxide, 2, 115

cardiologist, 121

cardiomyopathy, 125

cardiovascular disease, 133

carrier, 36, 37, 117, 128, 138, 201

cartilage, 51, 52, 54, 55, 64, 66, 161, 163, 207

cast, 20, 162, 170

casting, 100, 132

castration, 106

catalyst, 37

cataracts, 203, 204, 240, 242

Catholic, 190

cats, 17, 67, 176, 210, 231

cattle, 79

cave, 102, 108

cavities, 211

cell culture, 239

cell death, 59, 239

cell division, 12, 13, 21, 54, 199, 236, 240, 242,
    247

cell growth, 24, 92, 93, 121, 149, 235

cell surface, 144, 147, 157

central nervous system, 91, 95, 213, 237

centromere, 245

cerebellar ataxia, 40

cerebellum, 212, 213, 215

cerebrum, 189

CF gene, 117

channels, 117, 186, 208, 210

charged particle, 77

chemicals, 99, 132, 147, 178

chicken, 11

child abuse, 64, 145

childbirth, 137

childhood, 58, 66, 77, 98, 104, 116, 130, 140, 160,
    162, 164, 167, 177, 181, 183, 191, 200

childless, 37, 163

chimpanzees, 8

chloride, 78, 116, 117

chloroplasts, 3

cholera, 117

C

cabbage, 178

cholesterol, 144, 186, 222

chorea, 41

Christmas, 183, 228

chromatin, 48

chromosomal abnormalities, 13, 17, 20

chromosomes, 3, 5, 11, 13, 14, 16, 17, 18, 20, 21, 22, 23, 24, 25, 31, 37, 40, 42, 45, 46, 67, 203, 242, 245, 246, 247, 249

chronic disease, 186

chronic diseases, 186

chronic granulomatous disease, 152

cigarette smoke, 118

cilia, 119

CIPA, 95

circulation, 37, 115

circumcision, 142

cirrhosis, 186, 187, 207

classes, 80

classical, 154, 224

classroom, 208

cleaning, 151

cleft lip, 90, 141

cleft palate, 122, 139, 141

clone, 67

cloning, 67

closure, 54

clotting factors, 142, 143, 145, 151

clubbing, 123

clusters, 113

c-myc, 24

CNS, 213

coagulation, 142, 145, 186

coagulation factors, 186

coal mine, 202

coastal areas, 133

cocoon, 151

codes, 7, 75, 80, 145, 150, 190

coding, 12, 55, 64, 71, 72, 88, 90, 92, 178, 180, 184, 213, 222, 235, 239

codon, 248

codons, 245, 248

coeliac disease, 191

cognition, 41

cognitive function, 222, 231

cognitive impairment, 216, 220

cold war, 221

collagen, 51, 55, 62, 64, 71, 72, 144, 207

collisions, 3

colon, 237, 238

colon cancer, 237, 238

coma, 175

combined effect, 2

communities, 36, 60, 162, 179

community, 80

compatibility, 14

competence, 22

competition, 19

complement, 152

complement system, 152

complexity, 5, 7, 9, 11

complications, 72, 76, 134, 166, 191, 225

components, 2, 147, 176, 222

compounds, 174

concentration, 220

conception, 17, 44

concussion, 226

conduction, 124

configuration, 129

conflict, 211

confusion, 21, 175, 203, 249

congenital adrenal hyperplasia, 171

congenital erythropoietic porphyria, 135

congenital heart disease, 122, 123

congestive heart failure, 189

connective tissue, 56, 66, 71, 72, 242

consanguinity, 64

constipation, 136

construction, 42, 137

consumption, 175, 186, 231

contraceptives, 106

control, 23, 39, 47, 95, 133, 157, 160, 168, 203, 218, 219, 225, 226, 236, 242

control group, 133

conversion, 249

conviction, 27

convulsion, 190

copper, 111, 181, 187

coral, 201

coronary artery disease, 144

cortical bone, 65

corticosteroids, 166

cortisol, 160, 165

cotton, 183

coughing, 117

couples, 117

covering, 92, 108, 109, 116

cows, 67, 79, 231

crab, 60, 61, 235

craniofacial, 53

craniofacial dysostosis, 53

creativity, 224

cretinism, 164

Creutzfeldt-Jakob disease, 230

crime, 20, 27

criminal activity, 20

criminality, 21

criticism, 48, 179

cross-country, 200

cross-sectional, 110
crown, 211
crystals, 180
cues, 48, 212
cycles, 248
cycling, 142
cyclists, 142
Cysteine, 6
cystic fibrosis, 31, 42, 116, 151, 187
cysts, 92, 184
cytoplasm, 249
cytosine, 4, 48, 245, 248

D

daughter cells, 15, 236, 247
de novo, 52, 240
deafness, 64, 65, 207, 208, 209, 210, 211
deaths, 120, 123, 142, 186, 188
decay, 211
decisions, 136, 180
defence, 241
deficit, 122, 162
definition, 1, 5, 161, 204
deformities, 22, 29, 93
degenerative disease, 230
degradation, 64, 176, 223, 227
degrading, 180
dehydration, 116, 117
dehydrogenase, 131
delayed puberty, 158
delivery, 45
delusions, 229
dementia, 179, 220, 221, 222, 224, 225, 227, 228, 229, 230
demyelinating disease, 215, 216
demyelination, 216
dendrites, 213
Deng Xiaoping, 226
density, 62
dentistry, 128
deoxyribonucleic acid, 2
deoxyribose, 248
deported, 54
deposition, 65, 189, 224
deposits, 65, 66, 180, 187, 223, 227
depressed, 22, 128
depression, 135, 168, 189, 220
depth perception, 195
derivatives, 178
dermatitis, 106
dermis, 85, 89, 101, 105
desert, 36

destruction, 102, 118, 132, 148, 153, 183, 227, 233, 242
detection, 22, 210
developed countries, 176
diabetes, 48, 144, 159, 189, 240
diabetes insipidus, 159
diabetes mellitus, 159, 189
diaphysis, 65
diarrhoea, 117, 151, 175, 190, 191
diet, 48, 106, 135, 164, 177, 180, 242
dietary, 177, 178
diets, 177
differentiation, 121
digestion, 186, 190, 219
digestive tract, 178, 192
diploid, 247, 250
disabled, 179
discharges, 163
discomfort, 95
discrimination, 100
disease gene, 42, 179
diseases, 14, 29, 36, 37, 40, 51, 62, 64, 65, 77, 80, 82, 83, 90, 106, 117, 119, 130, 133, 141,
disequilibrium, 247
distraction, 243
distribution, 13, 168, 178
divergence, 246
division, 12, 13, 21, 54, 89, 92, 147, 199, 235, 236, 240, 242, 247
dizziness, 156, 211, 218
DNA, 2, 3, 4, 5, 6, 7, 8, 11, 12, 20, 23, 27, 29, 38, 39, 40, 41, 42, 47, 67, 75, 76, 96, 122, 129, 138, 144, 151, 180, 204, 235, 238, 239, 240, 241, 242, 245, 246, 247, 248, 249
DNA damage, 138, 236, 239, 240
DNA ligase, 239
DNA polymerase, 248
DNA repair, 236, 238, 239, 240, 242
DNA testing, 27
doctors, 74, 95, 128, 140, 161, 190, 215, 220, 225
dogs, 7, 80, 149, 194
domestication, 176
dominance, 30
dominant allele, 33
donkey, 46
donor, 67, 138, 183
dopamine, 226
dosage, 20
double helix, 4
Down syndrome, 15, 16
draft, 42
dream, 96
drinking, 146, 159, 161
Drosophila, 7

drugs, 80, 106, 134, 141, 149, 186, 242
drying, 106, 115, 149, 178
duplication, 22, 23, 246
duration, 202
dust, 65
dwarfism, 56
dyskinesia, 119
dyslexia, 228
dysplasia, 53, 55, 65, 90
dystrophin, 75

E

E. coli, 5
early retirement, 19
earth, 2, 85
eastern cultures, 59
Eastern Europe, 135, 179, 193
eating, 32, 175, 178, 216, 224, 232
Ebola, 6, 7
ectoderm, 90, 91, 101
eczema, 87
Eden, 212
egg, 2, 11, 38, 45, 67, 77, 138, 247, 250
Ehlers-Danlos syndrome, 71
elasticity, 55, 128
elastin, 71
elderly, 240, 242
election, 190
electrical conduction system, 124
electrical system, 124
embargo, 48
embryo, 45, 77, 90, 91, 119, 138, 139, 169, 232, 233
embryogenesis, 93
embryonic development, 60, 90, 101, 121, 192
embryonic stem, 67, 227, 232, 233
embryonic stem cells, 67, 232, 233
embryos, 13, 102, 138, 170, 233
emotional, 103, 210, 224
emphysema, 118
employees, 118
encephalopathy, 231
encoding, 6, 66, 121, 128, 215
endocrine, 157, 158, 159, 165
endocrine glands, 157
endocrine system, 157, 158
endoderm, 90
endorphins, 160
end-to-end, 11
endurance, 80
energy, 3, 38, 76, 131, 174, 179, 200, 203, 214, 226, 240
enlargement, 123, 163

enterprise, 217
environment, 2, 29, 31, 33, 48, 115, 150, 238
environmental factors, 103, 124, 247
enzymes, 42, 98, 111, 115, 133, 152, 171, 174, 176, 178, 179, 181, 239, 242
epidemic, 29
epidermis, 85, 87, 88, 89
epidermolysis bullosa, 88
epigenetic, 47
epigenetics, 47
epilepsy, 187, 188
epinephrine, 165
erythrocytes, 127
erythroid, 102
erythropoietin, 141
estradiol, 171
ethnic groups, 29, 202
etiquette, 162
eugenics, 37
eukaryote, 248
eukaryotes, 12, 246
eukaryotic cell, 3
evolution, 3, 29, 47, 48, 113, 117
examinations, 20, 125
excision, 239
exclusion, 37
excretion, 159, 178
execution, 58, 170
exercise, 125
exposure, 33, 48, 99, 104, 191, 238, 239
eye movement, 225

F

facial palsy, 65
factor VII, 142
factor VIII, 142
failure, 13, 59, 124, 149, 165, 170, 180, 181, 184, 189, 247
fainting, 78, 218
fairy tale, 143
familial, 32, 57, 78, 93, 111, 124, 180, 188, 204, 219, 222, 230
familial dysautonomia, 219
family history, 176, 197, 226
family life, 82
family members, 35, 103, 118, 211
famine, 48
farmers, 46
fasting, 134
fat, 2, 85, 168, 179
fatal familial insomnia, 230
fatigue, 78, 129, 131, 132, 138, 165, 184, 189, 191, 218

fats, 174, 179
feedback, 145
females, 12, 13, 16, 17, 18, 21, 36, 37, 38, 85, 102, 117, 164, 171, 191, 203, 249
fermentation, 3
fertility, 241
fertilization, 13, 45, 138
fever, 95, 118, 136, 151
fibrin, 142
fibroblast, 52, 242
fibroblast growth factor, 52
fibroblasts, 242
fibrosis, 31, 42, 116, 151, 187
filament, 125
film, 55, 88, 212, 216
films, 55, 82, 112
fingerprinting, 27, 246
fingerprints, 27
first generation, 36
First World, 61
fish, 86, 178
FISH, 22, 122
flatulence, 132
flexibility, 71
flooding, 133
flora, 190
flow, 60, 77, 119, 120, 123, 145, 155
fluctuations, 178
fluid, 65, 139, 183, 204, 211
fluorescence, 23
flushing, 237
focusing, 151, 196
folding, 11
folklore, 159
follicle, 105, 106, 107, 158, 160
follicle stimulating hormone, 158
follicles, 85, 96, 105, 106, 108
follicle-stimulating hormone, 160
food, 47, 48, 115, 116, 150, 173, 186, 190, 191, 200, 225
foodstuffs, 178
football, 107, 125, 140
founder effect, 41, 56, 66, 180, 204, 214
Fox, 226, 227
fractures, 51, 64
fragility, 64, 88
Franklin D. Roosevelt, 45, 220, 221
fraud, 67
free radical, 81
free radicals, 81
freedom, 44, 103
fresh water, 159
friction, 88
frog, 67

frontotemporal dementia, 224
frostbite, 96
fructose, 174
fruit flies, 11, 241
funding, 233
funds, vii, 225
fusion, 53, 250

G

gait, 218
games, 131, 200
gamete, 2, 247
gametes, 11, 13, 31, 240, 247, 249, 250
gaseous waste, 115
gases, 2
gastrointestinal, 118
gender, 139, 170
gene therapy, 151, 199
generation, 30, 32, 36, 40, 151, 245, 246, 247
genetic code, 245
genetic defect, 139, 147
genetic disease, 14, 29, 36, 37, 83, 151, 179, 189, 245, 249
genetic disorders, iv, vii, 90, 112, 184
genetic drift, 178
genetic factors, 106, 208, 222, 226
genetic information, 67, 246
genetic mutations, vii, 110, 162
genetic syndromes, 120
genetic testing, 139, 179
genetic traits, 139
genetics, 31, 47, 155
genome, 5, 6, 7, 27, 42, 47, 75, 203, 238, 247, 249
genomes, 6
genotype, 18, 20, 246, 248
Gerstmann-Sträussler-Scheinker syndrome, 230
gift, 108
gifted, 109
gigantism, 160
ginger, 17, 99
girls, 27, 139, 163, 169, 170, 171
gland, 154, 157, 158, 159, 160, 162, 163, 164, 165, 166, 237
glass, 62, 138, 199
glasses, 197
glaucoma, 204, 205
gloves, 103, 151
glucose, 131, 159, 165, 174, 242
glutamic acid, 7
glycine, 133
glycogen, 175, 187
glycolysis, 131
glycoprotein, 144

glycoproteins, 181
glycosylation, 242
goals, 140
gold, 20, 125, 140, 165
gonadotropin, 158
gout, 180
government, 61, 225
grains, 38
grapes, 45, 67
groups, 7, 29, 48, 125, 128, 133, 202, 248
growth factor, 52
growth hormone, 21, 52, 159, 160, 161, 162, 230
guanine, 4, 180, 245, 248
guidelines, 142
guilty, 100, 124

H

haemoglobin, 128, 130, 133, 186
hair cells, 211
hair follicle, 85, 96, 105, 106, 108
hair loss, 106, 107
hallucinations, 134, 136, 228, 229, 231
handling, 221
haploid, 247, 250
harbour, 203
hard matrix, 51
harm, 176, 229
head injuries, 226
health, 29, 48, 55, 116, 185, 187, 219
health problems, 48, 55
hearing, 15, 92, 119, 194, 207, 208, 209, 210, 211, 217, 230
hearing impairment, 208, 209
hearing loss, 92, 119, 194, 207, 209, 210
heart, 15, 72, 74, 75, 76, 95, 111, 119, 120, 121, 122, 123, 124, 125, 141, 144, 156, 164, 166, 186, 189, 210, 218, 219, 237, 240
heart attack, 124, 141, 144
heart disease, 120, 122, 123, 144, 240
heart failure, 124, 189
heart rate, 95, 218, 219
heart valves, 121
heartbeat, 124
heat, 85, 143
height, 21, 55, 64, 73, 85, 121, 161, 162
helix, 4
hematopoietic, 232
hematopoietic stem cell, 232
hematopoietic stem cells, 232
hemophilia, 143
hepatocytes, 186
heredity, 5
hermaphrodite, 169

heterochromia, 193, 194, 209
heterogeneity, 246
heterogeneous, 216
heterozygote, 31, 117, 128, 246
heterozygotes, 31, 32, 33, 37, 128
high blood pressure, 144, 185, 237
high risk, 88
high-speed, 42
Hippocrates, 169, 184, 204, 235
hips, 55, 95
histone, 245
HIV, 29, 100, 143, 147
HIV infection, 29
HLA, 153, 155
Holland, 48
homozygote, 31
hormone, 52, 104, 112, 141, 154, 157, 158, 159, 160, 161, 162, 163, 164, 167, 231
hormones, 20, 21, 106, 112, 134, 153, 157, 158, 159, 160, 164, 165, 169, 235
horse, 46, 59, 85
horses, 36, 67
hospital, 58, 124, 128, 129, 161, 165, 186
hospital stays, 129
hospitalization, 135
hospitals, 61
host, 3, 6, 67, 118, 148, 153, 249
host tissue, 118
hostility, 92, 237
hot water, 95
House, 154, 228
household, 37, 154
human, 2, 5, 7, 11, 12, 14, 27, 38, 42, 44, 51, 59, 67, 75, 77, 80, 85, 86, 89, 91, 102, 105, 106, 110, 131, 151, 177, 211, 214, 219, 230, 232, 233, 238, 242, 245, 249
human brain, 214
human embryonic stem cells, 67
human genome, 5, 42, 75, 238
Human Genome Project, 42
humans, 7, 9, 11, 17, 25, 27, 31, 59, 80, 117, 151, 171, 175, 186, 201, 208, 231, 233, 242, 249
hunting, 202
husband, 42
hybrid, 42
hybrid cell, 42
hybrid cells, 42
hybridization, 23, 122
hybrids, 42
hyperactivity, 89
hyperkeratosis, 86
hyperplasia, 171
hypertension, 220
hyperthyroidism, 154, 165

hypertrichosis, 108, 109, 135
hypertrophic cardiomyopathy, 125
hypnosis, 37
hypogonadism, 158
hypopigmentation, 103
hypoplasia, 54
hypothalamic, 158
hypothalamus, 157, 158, 160
hypothesis, 2, 3, 4, 155
hypothyroidism, 164
hypoventilation, 219, 228
hypoventilation syndrome, 219, 228

I

ice, 65, 87, 95
identical twins, 27, 155
identification, 40, 42
identity, 39
illusion, 112
imagination, 210
Immanuel Kant, 229
immigrants, 57, 131, 159
immortal, 233, 242, 243
immortality, 219
immune cells, 147, 153
immune memory, 149
immune response, 147, 149, 150, 153, 156
immune system, 89, 115, 147, 148, 149, 151, 152,
    153, 154, 155, 156, 160, 166, 187, 216
immunity, 147, 148
immunodeficiency, 149
immunoglobulins, 147
immunological, 150
immunotherapy, 149
impairments, 216, 233
impotence, 168
imprinting, 45, 46, 47
in situ, 23, 122
in situ hybridization, 23, 122
in utero, 102, 108
in vitro, 138
in vivo, 151
inactivation, 17, 249
inactive, 98, 113, 245
inbreeding, 35, 36, 38, 60, 64, 246
incest, 36
incidence, 16, 41, 56, 78, 92, 117, 122, 129, 133,
    135, 148, 149, 150, 177, 179, 189, 197, 210,
    214, 231
incisor, 91
incurable, 143, 207
indigenous, 163
industrial, 118

industry, 118
infancy, 218, 219
infant mortality, 128, 133
infant mortality rate, 133
infants, 116, 123, 124, 137, 149, 164, 177, 182,
    210
infection, 29, 58, 115, 138, 143, 147, 155, 230
infections, 48, 95, 116, 117, 118, 119, 124, 147,
    149, 150, 152, 153, 209, 210
infectious, 86, 149, 230, 249
infertile, 20, 169
inflammation, 29, 89, 190
inflammatory, 143
inflammatory response, 144
ingestion, 132, 200, 226
inheritance, 18, 20, 21, 29, 30, 31, 33, 35, 36, 37,
    38, 42, 47, 48, 65, 71, 103, 110, 153, 155, 173,
    174, 176, 193, 198, 199, 204, 216, 218, 224,
    236, 241, 246, 247, 248
inherited disorder, 35, 55, 64, 72, 95, 187
injections, 149
injuries, 66, 142, 226, 233
injury, iv, 66, 75, 142, 187, 233
inner cell mass, 233
inner ear, 207, 208, 210, 211
innocence, 60
inorganic salts, 115
insertion, 151
insight, 45
insomnia, 230, 231
inspiration, 22, 92, 219
instability, 187, 189
institutionalisation, 134
institutions, 209
insulation, 215
insulin, 151, 159, 166
intellectual development, 178
intelligence, 177
intensive care unit, 129
interphase, 245
interval, 124
intervention, 178
interview, 82
intestine, 12, 116, 117, 176, 190, 191
intimidating, 162
intraocular, 205
intraocular pressure, 205
inventions, 210
inversions, 25
iodine, 164
ion channels, 210
ions, 78, 117, 124, 165, 208
IQ, 22, 47
iris, 193, 194, 195

iron, 128, 133, 181, 187, 189
irritability, 1, 168, 184
irritable bowel syndrome, 190
island, 101, 201
isolation, 37
isoleucine, 178
Israel, 36, 54, 83
Italy, 108, 132
IVF, 138

J

James Watson, 4
jaundice, 132, 186, 203
jobs, 186
joint pain, 142, 161
joints, 55, 65, 71, 90, 95, 156, 180, 212
juries, 124
justice, 60

K

Kant, 229
karyotype, 11, 14, 17
karyotypes, 18
Kentucky, 121
keratin, 86, 88, 105, 111
keratinocyte, 85, 88, 89, 96
keratinocytes, 86, 96
kidney, 119, 183, 184, 185
kidney failure, 184
kidney stones, 183
kidneys, 141, 156, 159, 165, 175, 181, 183, 185,
    187, 218
killing, 6, 147, 223
Kinase, 78
King, 46, 57, 63, 67, 96, 108, 136, 137, 171, 184,
    204
kuru, 230

L

labour, 119
lactase, 175
lactation, 163
lactose, 175, 190, 191
lactose intolerance, 175, 190
lamellar, 87
land, 237
language, 22, 205, 224, 230
large intestine, 176, 191
larynx, 163
later life, 207

lateral sclerosis, 216
Latino, 217
laughter, 147
laws, 29, 30, 125
leadership, 73
learning, 121, 208, 217, 223
left ventricle, 120
legislation, 227
lens, 175, 196, 198, 202
lenses, 196, 202
leprosy, 103
lesions, 91, 92, 104, 239
leucine, 178
leukaemia, 151
leukemia, 138
leukocytes, 127
levodopa, 227
Lewy bodies, 228
libido, 167
life cycle, 128, 249
life expectancy, 83, 117, 217
life span, 133
life style, 238
lifespan, 78, 128, 143, 145, 218, 241, 242
life-threatening, 77, 87, 184, 190
lifetime, 14, 48, 148, 175, 205, 219, 237
Li-Fraumeni syndrome, 237
likelihood, 27, 56, 155, 230
limitation, 242
linear, 195
linkage, 20, 42, 246, 249
links, 124
lipid, 2, 179, 186
lipid metabolism, 186
lipids, 179, 181, 215, 222
liver, 12, 119, 133, 134, 151, 175, 181, 185, 186,
    187, 189, 200, 207
liver cells, 186
liver cirrhosis, 207
liver damage, 189
liver disease, 182, 186, 187, 189
liver enzymes, 133
location, 22, 23, 41, 42, 56, 119
locus, 245, 246, 247
long period, 117, 151, 205
longevity, 133, 241, 242
Los Angeles, 73, 210
losses, 117
Lou Gehrig's disease, 81
Louisiana, 36
love, 133, 219
lover, 94, 243
lung, 115, 116, 118, 149, 151
lung disease, 118, 149

lungs, 115, 116, 118, 119, 120, 156
lupus, 156, 207
luteinizing hormone, 160
lying, 117
lymphatic, 38, 85
lymphocytes, 67, 147, 150
lymphoma, 24, 151
lysis, 88, 181
lysosomes, 99, 181

M

macules, 104
mad cow disease, 231
magnetic, iv
mainstream, 209
maintenance, 112, 216, 239
malaria, 128, 129, 131, 132
males, 12, 14, 16, 20, 36, 42, 75, 102, 117, 119, 133, 142, 159, 164, 169, 180, 184, 191, 201, 203, 217, 249
malignant, 92, 235
malignant cells, 235
malnutrition, 134, 147
mammal, 45, 85
Mammalian, 127
mammals, 46, 67, 175
mandibular, 31, 33
manipulation, 242
manufacturing, 76
Mao Zedong, 226
mapping, 42, 247
Marfan syndrome, 72, 73
marriage, 35, 241
marrow, 127, 138, 141, 150, 151
mask, 78, 94, 156
maternal, 13, 35, 38, 39, 45, 46, 47, 67, 78
maternal age, 13
matrix, 51
meals, 175
meat, 80
medical student, 180
medication, 228
medications, 106, 167
medicine, 37, 233
meiosis, 12, 13, 15, 21, 47
melanin, 96, 97, 98, 99, 101, 103, 104, 111, 166, 189, 193, 195, 210
melanoma, 104
melanosomes, 96, 97, 99, 104
melody, 224
membranes, 179, 183, 241
memory, 15, 47, 148, 149, 150, 190, 222, 223, 224
memory loss, 224

men, 20, 21, 51, 91, 100, 106, 115, 125, 157, 164, 168, 201, 220, 226
Mendel, 29, 30, 67
meningitis, 153
menopause, 51
menstrual cycle, 154
menstruation, 21
mental ability, 17
mental capacity, 221
mental development, 175
mental retardation, 16, 41, 47, 165, 176, 177, 178, 180, 181
mesoderm, 90
messengers, 157
metabolic, 174, 216, 238, 241
metabolic disorder, 216
metabolic pathways, 174
metabolic rate, 241
metabolism, 1, 129, 164, 173, 174, 175, 176, 177, 178, 180, 186, 189, 222, 241
metabolite, 175
metabolizing, 175
metafemale, 18
metals, 187
metaphase, 245, 246
metastasis, 235
methionine, 7, 178
methyl group, 48
methyl groups, 48
mice, 7, 11, 48, 80, 112, 117
microcephaly, 16
microorganisms, 152
microscope, 1, 22
midbrain, 224
Middle Ages, 116
middle-aged, 111, 241
migraine, 197
migraines, 197
migration, 101
military, 78
milk, 163, 164, 175
minerals, 51
minority, 36, 71
miscarriage, 14
miscarriages, 14
mites, 106
mitochondria, 3, 38, 76, 181, 203, 214
mitochondrial, 39, 76, 203, 247
mitochondrial DNA, 39, 76, 204
mitosis, 12, 15, 21, 47, 235, 245, 248
mitotic, 247
mitral, 73
mitral valve, 73
mixing, 112

mobility, 66, 119

moisture, 86, 91

mole, 45

molecular structure, 4

molecules, 2, 4, 99, 119, 128, 133, 142, 147, 152, 174, 181, 208, 227, 231, 241, 245, 249

money, 65, 94, 103, 217, 233

monkeys, 65

monks, 1, 109

monosaccharides, 174

Monozygotic, 247

Monroe, 211

mortality, 6, 128, 132

mortality rate, 6, 132

mosquitoes, 128

mothers, 21, 42, 77, 139, 190, 203

motion, 211

motion sickness, 211

motor neuron disease, 81, 82, 83, 226

motor neurons, 81, 95, 213

mould spores, 124

mouse, 42, 48, 67, 208

mouth, 22, 78, 122, 141

movement, 41, 119, 210, 211, 213, 217, 218, 219, 225, 226, 230

mRNA, 245, 247, 248, 249

MS, 112, 155

mtDNA, 38, 40

mucin, 115

mucus, 115, 116, 118, 119, 159

multicellular organisms, 2, 3, 12

multiple alleles, 27

multiple sclerosis, 155, 216

multipotent, 232

murder, 20, 27, 60, 123

muscle, 6, 66, 75, 76, 77, 78, 79, 80, 81, 85, 95, 120, 125, 135, 155, 213, 216, 218, 219, 226, 233

muscle cells, 75, 95, 125, 155

muscle contraction, 78, 81

muscle strength, 79

muscle weakness, 75, 76, 78, 135, 155, 216, 218

muscles, 66, 75, 77, 80, 81, 95, 125, 166, 175, 210, 212, 213, 215, 218, 226

muscular dystrophy, 76, 80, 151

music, vii, 16, 22, 58, 162, 199, 205, 208

musicians, 22, 207

mutant, 35, 42, 53, 54, 77, 80, 87, 113, 117, 118, 124, 128, 130, 136, 151, 171, 178, 179, 208, 214, 225

Myasthenia Gravis, 155

Mycobacterium, 179

myelin, 155, 215, 216

myeloid, 102, 138

myopathies, 75, 76

myopathy, 77

myopia, 196, 197

myopic, 197

myotonic dystrophy, 40, 78

N

naming, 102

Nash, 138

nation, 96, 221

National Basketball Association, 73

natural, 7, 21, 49, 178

natural science, 49

natural selection, 178

nausea, 211, 237

navy, 146

neonatal, 83

nephritis, 31, 183

nerve, 2, 76, 86, 92, 95, 155, 191, 195, 198, 203, 204, 213, 215, 216, 242

nerve cells, 2, 76, 95, 191, 214, 215

nerves, 64, 65, 81, 85, 91, 92, 162, 213, 215, 217

nervous system, 91, 92, 95, 213, 218, 219, 237

network, 142

neurasthenia, 218

neurodegeneration, 219

neurodegenerative, 223, 225, 226

neurodegenerative disease, 223, 225

neurodegenerative diseases, 223, 225

neurodegenerative disorders, 223, 226

neurofibrillary tangles, 223

neurofibroma, 92

neurological condition, 92, 197

neurological disorder, 218

neurologist, 216

neuronal cells, 12

neuronal loss, 41

neurons, 2, 13, 41, 81, 95, 179, 181, 213, 216, 223, 226, 230

Neuropathies, 215

neuropathy, 203, 215

neurotoxic, 83

neurotoxins, 134

neurotransmitters, 155

New York Times, 77

New Zealand, 107, 180

NFL, 140

Nielsen, 210

nitrogen, 2

Nobel Prize, 4

nodes, 111

nodules, 66

non-profit, 210

norepinephrine, 165
normal aging, 222, 240
nuclear, 67, 245
nucleic acid, 2, 245, 248
nucleotide sequence, 27
nucleotides, 2, 40, 180, 245, 247, 248
nucleus, 2, 3, 5, 11, 38, 67, 77, 96, 157, 240, 245, 246, 247, 248, 249
nurse, 148
nurses, 21
nursing, 82
nursing care, 82
nutrients, 1, 186, 191
nutritional deficiencies, 191
nyctalopia, 199
nymph, 219
nystagmus, 195

ovary, 119
overeating, 47
overproduction, 86, 160
overweight, 112
ovulation, 163
ovum, 2, 119, 246
ownership, 241
oxidants, 132
oxidative, 132
oxidative damage, 132
oxygen, 3, 115, 120, 123, 128, 129, 133, 141, 152, 186, 236
oxygenation, 115
oxytocin, 158

O

oat, 78
obesity, 47, 48, 112, 228
obesity hypoventilation syndrome, 228
oestrogen, 21, 158, 169, 171
oil, 86, 179, 217
old age, 14, 108, 218, 240
older people, 222, 238
olfactory, 119
Olympic Games, 20, 165
oncogene, 236, 237, 248, 249
Oncogene, 248
oncogenes, 235, 236
onion, 5
oocyte, 2, 13, 31, 67
oocytes, 13, 21
oogenesis, 13
opposition, 44
optic nerve, 198, 203, 204
optimism, 221
oral, 106
oral contraceptives, 106
organ, 3, 56, 95, 186, 205
organelles, 3, 38, 99, 181
organic, 133
organism, 2, 5, 7, 11, 12, 30, 67, 147, 245, 246, 248
orientation, 210, 212
Osama bin Laden, 73
ossification, 66
osteogenesis imperfecta, 62, 71, 207
osteoporosis, 51, 167, 240
otosclerosis, 207
ovarian cancer, 4, 238
ovaries, 21, 157, 158

P

packets, 96, 104
pain, 71, 88, 90, 95, 129, 134, 137, 142, 155, 160, 161, 167, 176, 180, 190, 216
paints, 193
palate, 90, 141
palpitations, 237
pancreas, 116, 157, 189
Papua New Guinea, 230
paralysis, 79, 81, 179, 228
paranoia, 220
parasite, 128, 132
parathyroid, 157
parathyroid glands, 157
parent-child, 193
parents, 31, 32, 33, 35, 52, 57, 64, 88, 131, 142, 150, 159, 162, 190, 210, 216, 240
Parkinson, 223, 225, 226, 227, 228, 233
Parkinson disease, 225
particles, 78
pasta, 174
paternal, 35, 40, 45, 46, 53, 103
paternity, 27
pathogens, 118, 147, 149, 152, 153
pathways, 78, 148, 174, 180
patients, 64, 76, 95, 102, 106, 119, 134, 135, 143, 145, 149, 151, 155, 189, 197, 226, 231
patterning, 17, 60
PCR, 248
pedigree, 247
penetrance, 33, 134
peptide, 248
peptide bonds, 248
perception, 195, 197, 200, 201
perceptions, 59
performers, 54
periodic, 128, 211
peripheral nerve, 215

peripheral nervous system, 213
peroxisomes, 181
personality, 41, 48
pesticides, 83
phage, 152
phagocyte, 153
phenotype, 246, 247, 248
phenotypes, 61, 246
phenotypic, 48, 248
phenylalanine, 176, 177
phenylketonuria, 129, 176, 177, 178
phosphate, 4, 131, 248
photographs, 4, 33, 98, 167
photoreceptor, 198
photoreceptor cells, 198
photoreceptors, 201
photosensitivity, 135, 136, 137
photosynthesis, 3
physical activity, 73
physicians, 37, 95
physics, 82
physiopathology, iv
pig, 237
pitch, 22
pituitary, 157, 158, 159, 160, 161, 162, 163, 166,
    231
pituitary gland, 157, 158, 160, 162, 163, 166, 231
PKD, 184
placenta, 171
placental, 139
plague, 148
plants, 29, 30, 31, 67
plaques, 155, 223
plasma, 127, 152
plasma proteins, 152
plasmodium, 132
plastic, 150
platelet, 142, 143, 145
platelets, 99, 127, 138, 144, 232
play, 47, 58, 66, 72, 102, 115, 119, 143, 155, 191,
    193, 207, 223
pluripotency, 233
pneumonia, 66, 116, 119, 149, 221
poison, 146, 154, 173
police, 27, 185
political leaders, vii
politics, 132
pollutants, 118
polycystic kidney disease, 119, 184
polycythemia, 141
polydactyly, 56, 57, 139
polygenic, 31
polymerase, 248, 249
polymorphism, 42, 246

polymorphisms, 27, 246
polypeptide, 245, 246, 247, 248, 249
polypeptides, 247
polysaccharides, 174
poor, 93, 119, 161, 219
population, 16, 23, 27, 30, 36, 38, 41, 56, 57, 60,
    66, 89, 99, 102, 103, 113, 117, 139, 144, 145,
    153, 160, 175, 178, 179, 183, 190, 196, 197,
    201, 208, 222, 246, 247, 248
pore, 51
porphyria, 38, 134, 135, 136, 137
porphyrins, 133, 134
Portugal, 23
posture, 218
potassium, 51, 124, 210
power, 27, 76, 77, 80
Prader-Willi syndrome, 47
pregnancy, 21, 45, 108, 139, 150, 163, 171
pregnant, 14, 48, 163
prejudice, 53
preschool, 210
preschool children, 210
president, 40, 166, 184, 190, 220, 233
pressure, 86, 95, 144, 166, 185, 204, 219, 237
primaquine, 132
primary products, 249
primates, 242
prion diseases, 230, 231
prions, 230
private, 210
probability, 37, 124
probe, 23
profit, 210
program, 204
progressive supranuclear palsy, 224
prokaryotic, 2, 3
prokaryotic cell, 2, 3
prolactin, 160, 163
proliferation, 6, 24, 89, 235, 248, 249
property, iv, 4
proposition, 241
prosecutor, 27
prostrate, 17
protection, 86, 117, 131, 138, 209
protein function, 52, 66
Proteins, 66
proto-oncogene, 235, 236, 237
protozoa, 1
pseudo, 54
psoriasis, 89, 103
psychiatric patients, 134
psychiatrist, 220, 224
psychoses, 134, 220
psychosis, 134, 136, 168

ptosis, 78
puberty, 14, 17, 106, 158, 163, 169, 170, 171, 240
public, 82, 87, 92, 190, 221
pulse, 136
pumps, 120
punishment, 181
pupil, 23, 193, 194
pupils, 23
purines, 180
pus, 149
pyogenic, 149

Q

QT interval, 124

R

race, 7, 97, 209
racial groups, 125
radiation, 2, 96, 104, 236
radiation damage, 236
radius, 23
rain, 65
rancid, 178
random, 81, 178, 249
range, 15, 16, 149, 151, 178, 179, 191, 224, 238, 248
rape, 100
rash, 156
rat, 67, 146
rats, 242
reactive oxygen, 152
reading, 195, 200, 205, 211, 246, 248
reality, 109, 229
recall, 18
receptors, 95, 106, 112, 155, 235
recessive allele, 33
recognition, 44, 49, 88, 239
reconstructive surgery, 20
recovery, 136, 191, 203
red blood cell, 127, 130, 132, 232
red blood cells, 127, 130, 132, 232
redundancy, 153
refining, 195
regenerate, 186
regeneration, 203
regenerative capacity, 186
regenerative medicine, 233
regular, 21
regulation, 205
regulatory bodies, 125
relationship, 27, 238

relationships, 248
relatives, 27, 36, 204, 230, 238, 246
remodelling, 48
Renaissance, 41, 195
renal, 42, 180, 183
renal disease, 42
renal failure, 180
rent, 183
repair, 12, 138, 155, 236, 238, 239, 240, 242
replication, 3, 132
reproduction, 1, 13, 31, 38, 67, 157
reputation, 46
residential, 209
residues, 48
resistance, 29, 118, 128, 179
resources, 47
respiration, 3, 95, 164
respiratory, 76, 115, 116, 117, 119
respiratory problems, 117
restaurants, 203
retardation, 16, 41, 47, 165, 176, 177, 178, 180, 182
retina, 195, 196, 198, 199, 203, 210, 237
retinitis, 199
retinitis pigmentosa, 199
retinoblastoma, 198, 237
retinol, 199
retribution, 60
retrovirus, 6
reverse transcriptase, 249
Reynolds, 118
rhythm, 22, 124
ribonucleic acid, 2, 5
ribose, 248
ribosomes, 248, 249
right ventricle, 120
rigidity, 226
risk, 14, 82, 88, 95, 117, 125, 139, 141, 144, 153, 204, 222, 226, 238
risk factors, 82, 144, 153
risks, 15, 35, 58
Rita, 167, 168
RNA, 2, 3, 5, 6, 180, 245, 246, 247, 248, 249
Roman Empire, 187
royalties, vii
royalty, 32
rubella, 210

S

sadness, 135
saline, 11
salmonella, 118
salt, 116, 166

salts, 51, 115, 183
sample, 20, 27, 139
sampling, 139
scabies, 106
scaffolds, 86
scalp, 98, 101, 106, 107, 112, 170
scar tissue, 155
school, 170, 200, 209, 240
sclerosis, 66, 81, 91, 92, 93, 155, 216
search, 23, 88, 216
Second World, 48
Second World War, 48
secret, 184, 185, 225
Secret Service, 154
secrete, 117, 165
secretion, 157, 158
seed, 45
seeds, 29, 227
seizure, 125
seizures, 134, 175, 188
selecting, 80, 113
self-mutilation, 134
self-renewing, 233
semen, 20
senescence, 242
senile, 224
senile dementia, 224
sensation, 95, 135, 215
sensations, 95, 160, 215
sensitivity, 166, 195, 200, 239
sensory nerves, 215
separation, 13, 248
sequencing, 7, 42
series, ii, 4, 7, 16, 21, 55, 60, 155, 220, 239
services, iv, 210
settlers, 44, 180
severity, 62, 64, 71, 92, 145, 156, 180
sex, 12, 17, 18, 19, 27, 63, 100, 110, 153, 158,
    159, 167, 169, 170, 171, 245
sex chromosome, 12, 17, 19, 20, 245
sex hormones, 153
sex steroid, 169
sexual development, 158
sexual reproduction, 13, 38
shape, 2, 23, 44, 51, 65, 93, 110, 127, 128, 129,
    130, 156, 196, 198
sharing, 7, 56, 75, 108, 227
sheep, 67, 231
shock, 95, 124
shock waves, 124
shoot, 60
short term memory, 222
shortage, 48, 141
shortness of breath, 118

short-term, 15, 222
short-term memory, 15, 222
shoulder, 60
shoulders, 66
siblings, 162, 177, 184, 190
sickle cell, 42, 128, 129, 131, 151
Sickle cell, 128, 129, 130
SIDS, 123
sign, 73, 78, 164
signalling, 31, 52, 121, 184, 230
signals, 155, 157, 158, 198, 213, 215, 226, 235
signs, 48, 72, 92, 93, 138, 220, 221, 230
silver, 73
sites, 27, 29, 235
situs inversus, 119
skeletal muscle, 75, 80
skeleton, 51, 66, 94, 161
skills, 15, 109, 202
skin cancer, 88, 96, 104, 239
skin disorders, 86
sleep, 94, 219, 230
SMA, 83
small intestine, 116, 117, 191
smallpox, 29, 148
smoke, 115, 118
smoking, 118, 144
snoring, 219
soccer, 83, 140
social behaviour, 224
social security, 27
SOD1, 81
sodium, 116, 165, 183
soil, 164
soils, 164
solar, 104
somatic cell, 42, 67, 151, 247
somatic cell nuclear transfer, 67
somatic cells, 151, 247
somatic nervous system, 218
Sonic hedgehog, 102
sounds, 140
Soviet Union, 7, 220
spastic, 218
spasticity, 81, 179, 218
speciation, 25
species, 7, 11, 44, 46, 171, 231
specificity, 231
speculation, 170
speech, 15, 42, 79, 82, 136, 195, 217, 219, 221,
    224, 226
speed, 42, 80
sperm, 1, 2, 11, 12, 13, 14, 17, 20, 21, 31, 38, 45,
    119, 151, 246
spheres, 2

spherocytosis, 129

spin, 58

spina bifida, 139

spinal cord, 2, 81, 91, 92, 213, 215, 216, 218, 233

spinal muscular atrophy, 83

spine, 72, 74, 82, 92, 95

spinocerebellar ataxia, 214

spleen, 128, 129, 132

sporadic, 160, 211

sports, 73, 125

sprains, 66

springs, 85

stability, 88, 203

stages, 139, 220, 240

Star Wars, 55

starch, 223

stars, 140, 141, 199

starvation, 48

Stem cell, 232

stem cell research, 227, 233

stem cells, 67, 89, 127, 138, 139, 232, 233

Stem cells, 232

Stephen Hawking, 82, 83

sterile, 46

steroid, 169, 170

steroid hormone, 169

steroid hormones, 169

steroids, 157, 165, 166, 168, 171

stiffness, 78, 218

stomach, 176

storage, 175, 179, 181, 187

storms, 2

strabismus, 195

strain, 80

strength, 55, 77, 79, 80, 216

stress, 37, 95, 103, 107, 165, 210, 236

stressors, 236

stretching, 75, 135

stroke, 128, 133

strokes, 141, 220

structural defect, 14, 22

structural defects, 14, 22

structural protein, 71, 86, 87

students, 197

substances, 173, 174, 179, 181, 186, 215, 216

sucrose, 174

suffering, 48, 80, 90, 102, 124, 128, 132, 135, 136, 145, 181, 185, 188, 190, 201, 220, 222

sugar, 4, 132, 166, 174, 175, 181, 248

sugars, 174

suicide, 134, 146, 165, 170, 189

suicide attempts, 134

sulphur, 111, 178

summer, 48, 124, 202

sunlight, 96, 113, 137

superoxide, 242

superoxide dismutase, 242

superstitious, 109

supply, 51, 160

suppression, 163

suppressor, 93, 199, 235, 236

suppressors, 92, 93, 236

Supreme Court, 199

surface area, 85, 129

surgeons, 186

surgery, 17, 18, 20, 54, 141, 190, 207

surgical, 18

surprise, 29

survival, 132

surviving, 76, 83

susceptibility, 73, 118, 119, 133, 138, 149, 153, 155, 204, 216, 239, 242

suspects, 242

sweat, 86, 90, 176, 178

Sweden, 175, 202

swelling, 163, 167, 211, 235

symptom, 142, 180, 184, 199

synapses, 179

synovial fluid, 65

synthesis, 2, 98, 165, 247, 248

syphilis, 207

systems, 16, 18, 109, 116

T

T and C, 4

T cell, 148

T cells, 148

T lymphocyte, 148

T lymphocytes, 148

tangles, 223

taste, 116, 159

tau, 223, 224, 227

Tay-Sachs disease, 31, 179

T-cell, 122, 147, 149, 153

T-cells, 147, 149, 153

teachers, 197

technology, 42

teens, 159, 162, 240

telangiectasia, 239

telephone, 179

television, 16, 21, 59, 92, 141, 155

telomerase, 242

telomere, 242

telomeres, 242, 245

temperament, 46, 80, 190

temperature, 2, 95, 98, 115, 219

temporal, 224

temporal lobe, 224
tendons, 75, 95
testes, 17, 157, 158, 168, 169, 170
testimony, 227
testosterone, 17, 20, 106, 158, 163, 168, 169, 170, 171
Testosterone, 171
testosterone levels, 20
Tetralogy of Fallot, 122
thalassemia, 130, 131
Thalassemia, 131
therapy, 21, 140, 150, 151, 190, 198, 199
thinking, 38, 163, 228
threatening, 77, 87, 102, 184, 190
threshold, 40
throat, 154
thrombocytopenia, 144
thrombosis, 141, 144, 145
thrombotic, 145
thrombus, 145
thymidine, 239
thymine, 4, 245, 248
thyroid, 154, 157, 159, 163, 164, 165
thyroid gland, 154, 157, 159, 163, 164
timber, 237
tin, 216
tissue, 1, 23, 51, 66, 71, 72, 75, 81, 85, 90, 107, 118, 139, 154, 155, 156, 166, 182, 187, 195, 198, 213, 222, 230, 231, 233, 235
title, 55, 122, 197
titration, 201
T-lymphocytes, 147
tobacco, 118
tolerance, 176
tonic, 77
toxic, 174, 175, 176, 179, 181, 186, 216, 227
toxic effect, 176
toxic substances, 174, 181, 186
toxin, 117
toxins, 118, 154, 186, 226
trade, 88
tradition, 36
training, 124
traits, 37, 48, 139, 249
trans, 86, 131
transcriptase, 249
transcription, 5, 235, 248
transcription factor, 235
transfer, 67
transfusion, 143
transfusions, 143
translation, 5, 247, 248
translocation, 22
Translocations, 24

transmission, 47, 231
transmits, 157, 203
transplant, 150
transport, 216, 222
traps, 115
trauma, 82
travel, 82, 215
trees, 65
tremor, 226
trial, 20, 44, 58
tribes, 162
triggers, 226
trisomy, 13, 15, 16
trisomy 21, 15
tuberculosis, 179
Tuberculosis, 179
tuberous sclerosis, 91, 92
tubular, 183
tumour, 92, 93, 160, 161, 163, 168, 199, 235, 236
tumour suppressor genes, 235, 237
tumours, 91, 92, 93, 160, 163, 235, 237
turnover, 89
twin studies, 90
twins, 27, 155, 222, 246, 247
type 2 diabetes, 240
typhoid, 118
typhoid fever, 118
typhus, 207
tyrosine, 7, 98, 176
Tyrosine, 6

U

ultrasonic waves, 139
ultrasound, 139
ultraviolet, 2, 96
umbilical cord, 139
underproduction, 158, 160
undifferentiated cells, 232
urban areas, 179
uric acid, 180
urine, 133, 136, 159, 176, 177, 178, 183, 184
uterine cancer, 238
uterus, 45, 119, 140
UV, 236, 238, 239
UV exposure, 239
UV light, 238
UV radiation, 236

V

vaccination, 148
vaccine, 7

valine, 7, 178
variability, 96
variables, 2
variation, 2, 144, 145, 201, 248
vascular dementia, 220
vasopressin, 6, 158
vector, 151
vegetable oil, 179
vegetative reproduction, 67
Venezuela, 41, 42
ventricles, 120, 121
vertigo, 211
vessels, 21, 56, 73, 86, 120, 144, 156
victims, 100, 237
Victoria, 14, 37, 38, 94, 137, 142
village, 57, 60, 108
villus, 139
violence, 20
violent, 205
viral infection, 115
virus, 5, 6, 7, 29, 149, 151, 198
viruses, 2, 3, 6, 143, 151, 249
viscosity, 141
visible, 2, 105, 201, 248
vision, 15, 156, 161, 193, 195, 196, 197, 198, 199,
    200, 201, 202, 205, 210, 225
visual field, 199
visual perception, 197
vitamin A, 199
vitamin D, 96, 113
vitamins, 242
vitiligo, 102, 103
voice, 82, 161, 171, 240
voiding, 135
volleyball, 73
vomiting, 151, 176, 190, 219, 237
Von Recklinghausen's disease, 92
voters, 166

W

waking, 219
walking, 42, 76, 213, 218, 228
war, 61, 220, 221
warfare, 7
warfarin, 145
waste disposal, 181
waste products, 183
wastes, 115
water, 2, 87, 95, 115, 117, 127, 154, 159, 165,
    211, 218, 219
water supplies, 159

water vapour, 2
weakness, 75, 76, 78, 79, 81, 83, 135, 155, 156,
    175, 183, 216, 218
weapons, 221
wear, 55, 103, 140, 197, 199
weight gain, 116
weight loss, 117, 165, 190, 191
Weinberg, 238
western culture, 59
Western Europe, 176
wheat, 191
wheelchair, 76, 156
white blood cells, 29, 118, 127, 138, 151, 232
White House, 154
wild animals, 108
William James, 185
winning, 84, 177, 208, 227
wisdom, 59
witchcraft, 58
women, 18, 19, 21, 37, 38, 44, 48, 51, 60, 71, 100,
    106, 112, 123, 140, 156, 157, 163, 168, 170,
    201, 204, 218
wood, 1, 66
wool, 97, 105
word of mouth, 21
workplace, 119
World Health Organisation, 104
writing, 61, 77, 135, 145, 229

X

X chromosome, 12, 15, 16, 17, 18, 20, 21, 37, 41,
    42, 75, 90, 101, 106, 132, 142, 149, 150, 159,
    169, 184, 201, 245, 249
xeroderma pigmentosum, 239
X-linked, 37, 42, 75, 90, 101, 142, 149, 180, 217,
    218
X-rays, 51, 64
XYY syndrome, 20

Y

Y chromosome, 12, 14, 16, 17, 18, 20, 40, 110,
    169, 249
yeast, 7, 245

Z

zygote, 31, 247